Essential
Electronic Design Automation (EDA)

www.phptr.com

Library of Congress Cataloging-in-Publication Data available

Editorial/production supervision: *Kathleen M. Caren*
Acqusition Editor: *Bernard Goodwin*
Editorial Assistant: *Michelle Vincenti*
Marketing Manager: *Dan DePasquale*
Manufacturing Manager: *Maura Zaldivar*
Cover Design Director: *Jerry Votta*
Interior Design: *Gail Cocker-Bogusz*

Publishing as Prentice Hall Professional Technical Reference
Upper Saddle River, NJ 07458

First Printing

ISBN 0-13-182829-0

Pearson Education LTD.
Pearson Education Australia PTY, Limited
Pearson Education Singapore, Pte. Ltd.
Pearson Education North Asia Ltd.
Pearson Education Canada, Ltd.
Pearson Educación de Mexico, S.A. de C.V.
Pearson Education—Japan
Pearson Education Malaysia, Pte. Ltd.

Prentice Hall Modern Semiconductor Design Series

James R. Armstrong and F. Gail Gray
VHDL Design Representation and Synthesis

Mark Gordon Arnold
Verilog Digital Computer Design: Algorithms into Hardware

Jayaram Bhasker
A VHDL Primer, Third Edition

Mark D. Birnbaum
Essential Electronic Design Automation (EDA)

Eric Bogatin
Signal Integrity: Simplified

Douglas Brooks
Signal Integrity Issues and Printed Circuit Board Design

Kanad Chakraborty and Pinaki Mazumder
Fault-Tolerance and Reliability Techniques for High-Density Random-Access Memories

Ken Coffman
Real World FPGA Design with Verilog

Alfred Crouch
Design-for-Test for Digital IC's and Embedded Core Systems

Daniel P. Foty
MOSFET Modeling with SPICE: Principles and Practice

Nigel Horspool and Peter Gorman
The ASIC Handbook

Howard Johnson and Martin Graham
High-Speed Digital Design: A Handbook of Black Magic

Howard Johnson and Martin Graham
High-Speed Signal Propagation: Advanced Black Magic

Pinaki Mazumder and Elizabeth Rudnick
Genetic Algorithms for VLSI Design, Layout, and Test Automation

Farzad Nekoogar and Faranak Nekoogar
From ASICs to SOCs: A Practical Approach

Farzad Nekoogar
Timing Verification of Application-Specific Integrated Circuits (ASICs)

Samir Palnitkar
Design Verification with **e**

David Pellerin and Douglas Taylor
VHDL Made Easy!

Christopher T. Robertson
Printed Circuit Board Designer's Reference: Basics

Samir S. Rofail and Kiat-Seng Yeo
Low-Voltage Low-Power Digital BiCMOS Circuits: Circuit Design,Comparative Study, and Sensitivity Analysis

Frank Scarpino
VHDL and AHDL Digital System Implementation

Wayne Wolf
Modern VLSI Design: System-on-Chip Design, Third Edition

Kiat-Seng Yeo, Samir S. Rofail, and Wang-Ling Goh
CMOS/BiCMOS ULSI: Low Voltage, Low Power

Brian Young
Digital Signal Integrity: Modeling and Simulation with Interconnects and Packages

Bob Zeidman
Verilog Designer's Library

Contents

Preface

PURPOSE OF THIS BOOK

The tremendous increase in the use of tiny electronic devices is common knowledge. We find them everywhere today, in cars, household appliances, telephones, music, and business equipment. The typical car or house uses dozens of them.

These devices are called microchips or *integrated circuits (ICs)*. Today a single IC can do more than an entire roomful of equipment just a decade ago. Integrated circuits are small enough to hold in your hand, yet contain millions of tiny electronic components.

Engineers create detailed design plans to make ICs, similar to an architect's building plans. Architects use computer tools to design a building and predict the structure's response to storms or earthquakes. Similarly, IC designers use computer program tools to design an IC, test its performance, and verify its behavior. We refer to the tools as *electronic design automation (EDA)*.

An entire industry has evolved to provide these tools to aid the IC designers. This book introduces readers to the EDA industry. It discusses both the technical and business aspects of EDA in clear non-technical language without equations. The text briefly describes the related semiconductor industry issues and evolving chip design problems addressed by the EDA tools. A unique, dialog format presents the technical material in an easy-to-read manner.

The book focuses solely on EDA for IC design, intentionally excluding other design automation areas (e.g., printed circuit boards and mechanical design). The text gives generic tool descriptions since company and product names change rapidly.

Intended Audience

The electronic product industry consists of electronic system manufacturers, semiconductor companies, and chip design houses. Semiconductor equipment providers, test equipment manufacturers, and EDA companies are also part of the industry.

In most of these firms, **over half** the employees are **non-technical** or "semi-technical." These semi-technical people are involved in the EDA or related industries. Experienced employees will have picked up some jargon and knowledge, but both they and most new employees lack an overall introduction to this highly technical field.

Sales and marketing, communications, legal, or finance personnel will find the book useful. Others in financial analysis, public relations, or publications firms also need to know about the EDA industry. Some readers will be interested only in the overview, business, or industry sections, while others will focus on particular technical EDA chapters.

Along with the **semi-technical** people, many people with **technical** backgrounds will find the book very beneficial. The technical backgrounds include computer engineering, programming, electronic testing, mechanical engineering, packaging, or academic fields. These readers may not have EDA backgrounds and so seek a simple introduction to EDA.

The book is thus helpful to new employees, both technical and non-technical. Some readers may be familiar with a portion of EDA and want to see "the big picture." Others may focus on technical areas relevant to their own work.

Faculty and **students** in universities, colleges, community colleges, and technical institutes can use the book as an introduction to the IC and EDA industries. The book will fit well in cross-discipline business/engineering courses. Technical students will find the full coverage useful and complementary to an academic course on ICs or EDA.

Non-technical readers include:

Within the organization:

Marketing communications, sales, and marketing personnel

Human relations, administrative personnel, and new hires

Manufacturing, purchasing, and operations personnel

Finance, accounting, and legal personnel

Outside the organization:

Financial analysts, law firms

Public relations, publications, or media representatives

Manufacturing representatives, personnel recruiters, or technical writers

Technical readers include:

Electrical engineers new to EDA

Mechanical, packaging, and quality assurance engineers

Programmers (software engineers)

Technical marketing and support personnel

Academic fields include:

Electrical, Mechanical, Systems, and Computer Engineering

Physics

Computer Science and Programming

Business, Marketing, and Management

Organization

The book's successive chapters build on each other, forming a logical sequence. However, most chapters can be read independently. The book may also serve as a reference source, using the several appendixes.

Chapter 1 gives an overview of EDA tools, the people who use them, and the design tasks they support. (EDA tools address specific design issues, so one has to understand those problems.)

Chapter 2 describes the EDA business itself. Chapter 3 provides a user perspective on EDA technical and business issues. Chapter 4 discusses the range of EDA tools and introduces some essential concepts.

Chapters 5, 6, and 7 focus on the three major EDA design tool areas: electronic system-level, functional chip-level, and physical. System-level tools help decide what the IC will do and how it will be made. Chip-level tools help design how the IC will operate (*function*). Physical design tools help implement the actual IC physical layout.

Chapter 8 discusses EDA industry trends and related IC design issues.

Since readers have a wide range of backgrounds, several appendixes fill in the technology basics. Appendixes A, B, and C introduce (in simple English) elementary electricity, semiconductor manufacture, and computer basics.

Many technical EDA and semiconductor terms are confusing. Most terms are metrics—each with different units of measure (such as inches, mils, or microns). Some are in English units, some are in metric units, and some are in both, depending on the context. Appendix D describes and compares these metrics.

Appendix E has pointers to other EDA reference sources for the reader to explore further. These include organizations, conferences, magazines, the Internet, and universities.

Appendix F provides more depth in several areas that affect the EDA business. These include deep submicron issues, intellectual property, and system-on-chip.

Every human enterprise (such as medicine, law, or academia) has its own jargon. EDA is no exception. There is a myriad of strange terms. Many come from the semiconductor world addressed by the EDA tools. In addition, there are all sorts of abbreviations and acronyms. The text defines many terms in context, and Appendix G provides an extensive glossary/acronym list with acronym pronunciation.

In summary, readers will be introduced to both the business and technical aspects of the EDA industry. They will learn about EDA tools, the designers who use them, and their design problems. In addition, they will gain insight into the current and future role of EDA in the electronics industries.

Acknowledgements

I thank my wife, Sarah, for her endless patience and support, and Lance Leventhal for his similarly endless reviews and insightful comments. My thanks also for the feedback from additional reviewers Gary Smith, Jim Turley, Jay Michlin, Patrick Kane, Rob Smith, and Yume Eng.

1 Introduction to EDA

In this chapter...

- Introduction
- EDA Party—Users and Tools
- EDA Benefits
- Summary

INTRODUCTION

Electronic Products

Engineers use *Electronic Design Automation (EDA)* tools to design electronic products. Electronic products include just about anything that plugs into the wall or uses a battery for electric power, such as computers, cell phones, digital cameras, and communications equipment. Electronics are used in houses, automobiles, aerospace products, and all kinds of industrial products.

To understand EDA we have to look more closely at electronic products. Like any technical area, electronics and EDA use lots of jargon. Let us begin with a little vocabulary so everyone can start with the same basic terms.

Printed Circuit (PC) Boards

Look inside a stereo, personal computer, or cellular phone, and you will see thin plastic (often green) printed circuit boards (*PC boards*).

Did You Know?

Printed Circuit (PC) boards are NOT the same as Personal Computers (PCs), and are NOT Politically Correct (PC). Unfortunately, the same abbreviation is used (PC). However, Personal Computers do contain PC boards.

Thin copper wires connect many little electronic parts mounted on the boards. These parts are small rectangular blocks (often black) with *pins* that stick out and look like insects with legs. The electronic parts come in different sizes and some are called *integrated circuits (ICs).* Figure 1.1 shows a drawing of the PC board, wires, and parts. The pins connect the electronic parts to the wires. Also note that the ICs may connect with pins as shown or with flat wires on all four sides, or with tiny solder balls underneath.

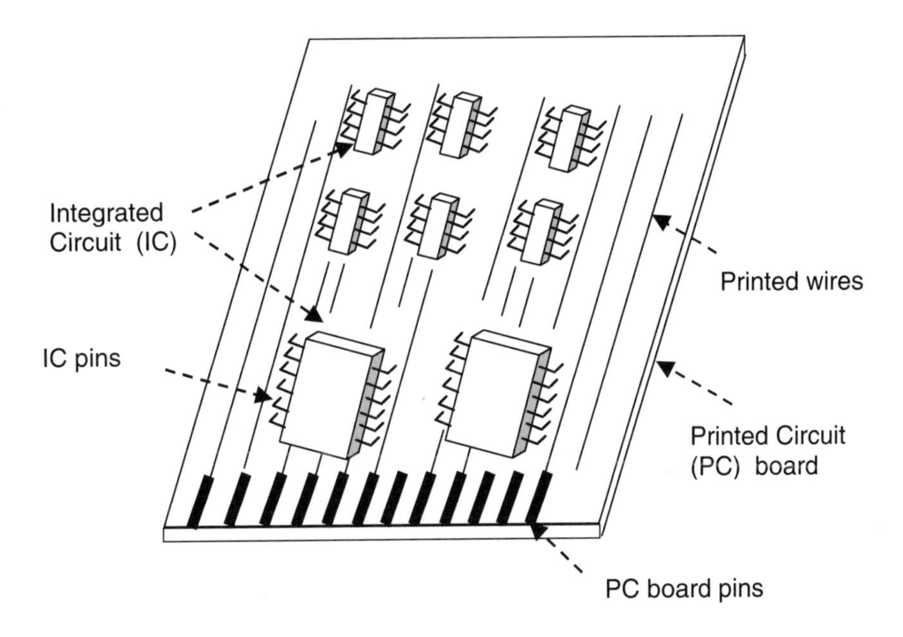

Figure 1.1
Printed Circuit Board

Note that the PC board wires are *printed* or deposited, and so are flat, not round. The PC board also has printed copper *fingers* or *connectors* at the edge for electrical connections off the board.

Integrated Circuits

Integrated circuits (ICs) use printed wiring very similar to that on the PC boards. The "board," however, is now a thin *silicon* chip, with **much** smaller devices and wiring. Both the devices and the wiring are fabricated in the silicon surface.

Did You Know?

The *silicon* (silly-con) here is a silver, brittle, metallic substance. It is a major element in ordinary sand. Do not confuse it with *silicone* (silly-cone), which is the rubbery material used in caulking and car waxes.

Semiconductor companies make ICs. These are also called chips, microchips, or silicon chips. Figure 1.2 shows a drawing of an integrated circuit, with its silicon chip, package, and pins.

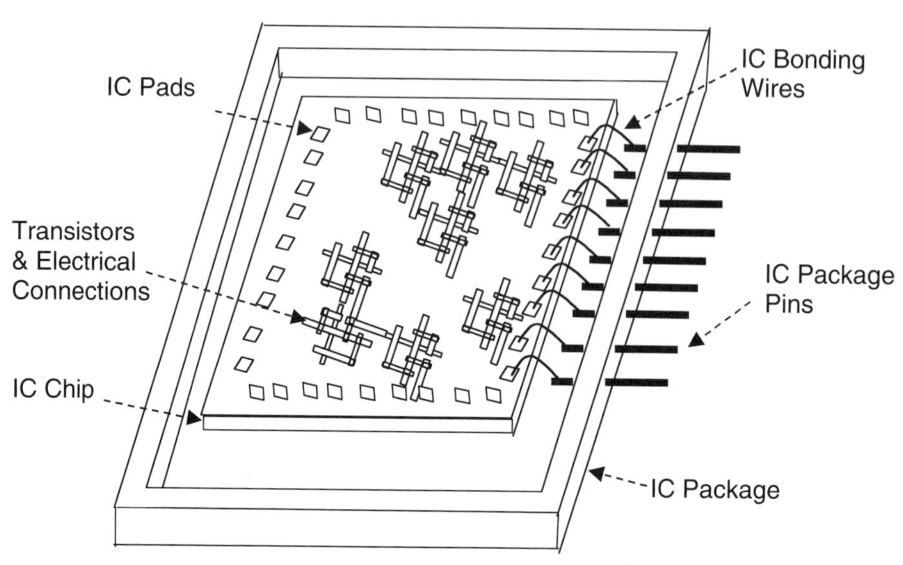

Figure 1.2
Integrated Circuit

The figure shows the fine gold *bonding* wires which connect the chip *pads* to the package *pins*. The wires are thinner than a human hair. The IC *package* takes up much more room than the tiny chip. There are many other styles of IC packages besides the example shown. Some chips are encased in smaller *chip scale* packages. Some ICs (called *flip chips*) can connect directly to PC boards, without package, wires, or pins. This allows those boards to hold more chips.

The ICs hold millions of tiny electrical switches called *transistors*. Thread-like printed wires on the IC connect transistors (and other devices) to form *electrical circuits*. These circuits are the heart of all electronic products. They can make small electrical signals larger (*amplify*) or make logical decisions (e.g., does number A = number B?). Basic logic circuits are called *gates*.

CAD, CAM, CAE, and EDA

Arranging the ICs and wire routes on the PC board is called *layout*. Programmers developed *Computer Aided Design (CAD)* software tools to help with the tedious PC board layout. Engineers later adapted PC board CAD tools for similar use on ICs.

Programmers continued to develop many other software tools to help design the IC and verify its behavior. Some are called *Computer Aided Engineering* (CAE). Others are called *Computer Aided Manufacturing (CAM)*. *Electronic Design Automation* (EDA) is an umbrella term for all these tools.

Data, Signals, and Input/Output

The information that is transferred between electronic products and ICs is called *data*. Data consists of numbers, letters, voice, video, etc. When electricity is sent from one IC to another, it is called a *signal.* Signals going into a product or IC are *inputs,* and those coming out are called *outputs.* Input and output together are referred to as input/output or *I/O.*

(You can read more about silicon, semiconductors, and computers in Appendices A, B, and C.)

Electronic Product Development

Neither the electronic products nor the ICs could be made without the use of EDA tools. EDA is intimately bound to the semiconductor IC and electronic product design industries.

Engineers use EDA tools to design electronic systems and ICs. To learn about EDA, we have to understand what the engineers are trying to do. Figure 1.3 gives an overall view from electronic product to IC.

In Figure 1.3, we see the *system engineers* discussing the idea for a new cellular telephone product. They create a set of IC requirements for the electronics which they need in the product. The requirements are similar to an architect's drawings.

Then IC *logic designers* transform those requirements into detailed design plans for the electronics. These are similar to the detailed blueprints for a house.

Next, IC *layout designers* take the design plans and lay out the physical view of the IC chip. For a house, this is like ordering real lumber, pipes, and appliances, with plans for the contractors.

The layout designers generate a computer data file for manufacturing. An IC fabrication plant uses this data file to make the chip. The product manufacturer assembles the resultant IC into the final cell phone product for use by a customer.

Let us now describe the companies, people, jobs, and tasks involved.

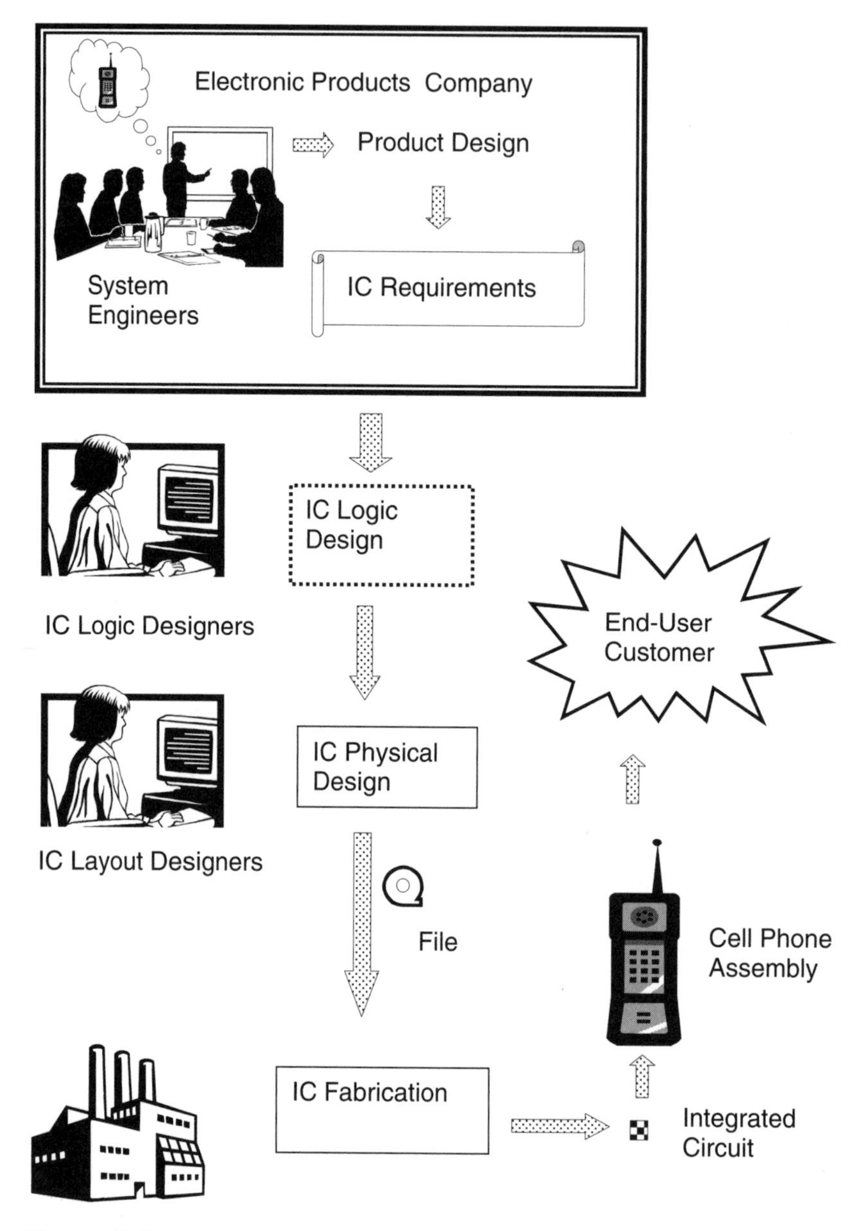

Figure 1.3
Electronic Product Development

EDA PARTY—USERS AND TOOLS

Nora Newbie is attending a party hosted by her new employer, an EDA company, Sandbox Tools, Inc. Sandbox is celebrating the release of a new company product, and several customers are at the party.

Nora is a recent college graduate with some public relations and marketing background. Let us listen in as she meets some new people. Her supervisor has just introduced her to Donna, an EDA tool developer with Sandbox.

Donna: Hi, Nora, welcome to Sandbox Tools.

Nora: Thanks, Donna. This is all new to me. I am not even sure what the company does. What do you do here?

Donna: The company develops EDA tools that engineers use to design ICs. I help write the programs, test them, and help our first customers try them out.

They suggest corrections or improvements that I make to the program. Then I retest the program, and the users try it again, until we get it right.

Nora: So, are the EDA tools similar to my word processor?

Donna: Somewhat, except they are specialized for electronic hardware design, instead of document creation, and with lots more data. EDA tools help engineers enter design ideas, verify that the ideas work, and check for errors. With millions of transistors on a chip, there is a lot of room for error.

Nora: The EDA tools must really be important. Did you work on this new product?

Donna: Yes, I put in many long hours. It is very satisfying to see the program finally working and getting used by the engineers.

Nora: Which engineers use our tools?

Donna: Engineers design ICs for communications equipment, industrial controls, military systems, and consumer electronics. EDA tools are for primarily system engineers, logic designers, verification engineers, ASIC designers, and layout designers.

Nora: I didn't realize there were so many types of engineers. Are they really all our customers?

Donna: Yes, and those are just some of the electronics engineering disciplines.

Nora: What do they all do?

Donna: Well, some of them are here at the party. Why don't I introduce you to them and they can tell you themselves. They all use EDA tools.

Nora: That sounds good to me.

System Design

Donna: Hey, Sam. This is Nora, who just joined Sandbox in the PR and marketing group. Can you tell her a little about what you system engineers do?

Sam: All right, Donna. Hello, Nora. Well, I work for SysComInc, a systems company that makes communication equipment for the Internet.

System engineers are like architects who provide a set of plans to a construction company. We help determine a product's requirements, explore ways of making it, and define all needed components. Then we make a model or prototype of the design. ICs are part of electronic products, so we end up specifying or designing those as well.

Nora: How does an EDA company like Sandbox Tools relate to all that?

Sam: Sandbox makes a *system modeling* tool which we use to check out our ideas. **Modeling** is a little like those real estate virtual tours on the Internet, where you can "walk" through a house. We model the riskiest parts of the design using your software tool. It saves us months of building a hardware prototype design.

In communications, many people may try to access a particular website at the same time. If our product takes too long to handle each access request, the customers go somewhere else. So we model the network loading (how many customers we can support per second). That tells us if our design approach will work.

Nora: The tools can predict if it will work or not?

Sam: Most of the time. Some parts of the design are complicated and hard to prove out any other way. It is fun to design things that have never been done before or were previously impossible. When you finally get your product to work, it is very satisfying.

Nora: Yes, Donna said that about her work, too. Do you design ICs as well?

Sam: I don't, but I sometimes describe what capabilities a chip must have for a specific product. Let me introduce you to Luigi, who is a logic designer, and Andrea, an ASIC engineer. They can explain about ICs better than I can.

Logic Design

Nora: Thanks, Sam. It's nice meeting you. Hi, Luigi.

Luigi: Hello, Nora. What can I do for you?

Nora: I'm new to Sandbox Tools, and I understand we make EDA tools for IC design. Can you tell me about that and what you do?

Luigi: Sure, Nora. I am a logic designer, and I work for Federal Semiconductor, a company that makes ICs. We are one of your customers. What we do is take the IC requirements from a customer or system designer like Sam. Then I design logic which implements the operations that he needs the IC to do.

Nora: What do you mean exactly by ***logic***?

Luigi: Well, for example, supposing I want to start my car.

A. I have to have the key, **AND**

B. The car has to have gas, **AND**

C. The car has to be in "Park."

If all of these things are true, the result is that I can start the car. If **any one** is false, then I can't start the car.

Think of it as a "decision box" with three inputs: A, B, and C. Think of the decision box having an output D, which is being able to start the car. D is true only if A **AND** B **AND** C are true. If any one of them (A, B, or C) is false, then the output will be false. That's logic.

Nora: That's all there is to it?

Luigi: Well, there are a few types of logical decisions. For example, supposing you, Sam, and Donna all have keys to my car. Then Nora **OR** Sam **OR** Donna could drive the car.

Think of another decision box—again with A, B, and C inputs. A is Nora being at the car, B is Sam at the car, and C is Donna at the car. This time, the output result, D, is that the car can be driven. D is true if any one of the inputs A **OR** B **OR** C is true.

We can make electrical circuits that make decisions just like the boxes. These logic circuits are called *gates.* What I just described was a logical **AND** gate, and a logical **OR** gate.

We can also make circuits that remember a value, whether it was true or false. Think of a light switch that you switch on or off. It stays where you last set it, either on or off. We can make *memory circuits* that behave the same way.

Most system problems can be described as a sequence of logical decisions. We can implement those with logic and memory circuits. I design the logic so that the inputs are transformed into outputs as specified.

Nora: So what do you end up with?

Luigi: A list of different logic gates, memory circuits, and the wires (or *nets*) that connect them. It's called a *netlist*.

Nora: Okay, I follow that—but where does the IC come in?

Luigi: The IC is full of transistors used to make logic gates and memory circuits.

Nora: So you choose the logic gates and the connections—is that logic design? Do you use any EDA tools?

Luigi: Oh yes. There are EDA tools to help the poor logic designer. We have tools to enter the design ideas into a computer using pictures (*graphic symbols*) or words (*text*). We describe the logic gates and the connections. There are various checking tools to catch entry errors.

We also have tools to verify that the logic does what we expected. Verification typically takes over two-thirds of the project design time. It is *iterative*—design, verify, re-design, verify, and so on. However, the Sandbox people have some new verification tools that should speed things up.

Nora: There seems to be more verification than design.

Luigi: It can be tedious, but it's like being a writer or artist—when I finish it—getting it to work—I feel a real sense of accomplishment.

Nora: So how does your netlist become an IC?

Luigi: You catch on quickly. You have noticed that this whole design process is a sequence of steps from idea to chip. Andrea, can you explain to Nora how my netlist becomes an IC?

ASIC Design

Andrea: Sure, Luigi. Hi, Nora. I am an ASIC engineer, and work for Fabless Design Co., an IC design company. We design ICs, but use an external foundry to manufacture our chips.

Nora: Hello, Andrea. What is an *ASIC* and what does an *ASIC* engineer do?

Andrea: Well, suppose you build a **custom** house where every part (door, window, etc.) is custom-made for you. Another approach is an architecture that uses standard-sized, prefabricated doors, or windows. With this **semi-custom** architecture, you can get all its parts from a standard catalog. So you still have some choice, but not the total flexibility of full custom. However, the semi-custom approach is much faster.

ASIC stands for *Application Specific Integrated Circuit*. ASICs are ICs designed using a semi-custom architecture. The architecture is an array of standard-sized circuits. A shorter design time is possible using *libraries* of standard-sized predesigned circuits (*gates or cells*) that fit into the array. A shorter design time gets to market faster, which everyone wants. There are several kinds of semi-custom ASIC architectures.

Nora: So what exactly do you do?

Andrea: In the house analogy, a paper plan may call for a 30-inch window. Then the architect would select a real physical window from the catalog (for example, ClearPane model 53, birch, dual pane, tinted glass).

For the IC, I design the logic using standard parts from libraries of gates or cells. The *logic design* is still like a set of architect's detailed plans at this point.

After completing the logic design, I transform it into a chip design that can be manufactured. Transforming (or mapping) the logic design to a real library of gates and memory elements is called *synthesis*. I can do this tedious job manually or with an automatic EDA synthesis tool.

The tool helps optimize the design for chip area, power, performance, and so forth. After running synthesis, I check to make sure that the logic has not changed and that all electrical rules are met.

After that, I turn the design netlist over to a layout designer. Let me find someone to explain layout to you.

Physical Layout Design

Let's see, I think Larry is here, and he is a layout designer. Hey, Larry. This is Nora, who just joined Sandbox. Can you please explain to her how you do your magic?

Larry: Sure. Hello, Nora.

Nora: Nice to meet you, Larry. Thanks, Andrea.

Larry: I work for Fabless Design Co., same as Andrea.

After Andrea has a netlist with real physical information, I do physical design steps called layout. These steps include physically locating or placing the many logic elements on the chip. Then the wiring between logic elements must be planned or routed.

Layout is all about minimizing chip area and interconnect wire lengths. The time delays through the gates and wires limit the speed of the chip. The shorter the wires, the less the delay and the faster the chip. The chip may be too slow due to a critical signal with a long interconnect wire.

Nora: How much delay are we talking about?

Larry: Nanoseconds... that's a billionth of a second. Light travels very fast. It could circle the earth eight times in a second. It can travel about one foot in a *nanosecond.* A personal computer typically executes one instruction in less than a nanosecond. So on a half-inch IC, we are worrying about fractions of nanoseconds. Electrons don't travel at the speed of light in wires, but they are pretty fast.

Nora: These are really short times...

Larry: Very short indeed. In addition, as chip features become smaller, the time delays become even more critical. The layout of the wires is very important. And now with things closer together on the chip, layout engineers have to worry about many new factors besides area and timing.

Layout consists of *placement* and *routing.* Using an EDA placement tool, we place the gates or cells at locations on the chip. Those cells with the most connections between them are placed close together.

A good placement enables short wires between the cells. It is similar to routing water pipes in a home. You want the water pipes to have the shortest path from the heater to the faucet, so you get hot water quickly.

After all the cells are placed, another EDA tool routes the wire connections between the cells. Sometimes I have to change the placement repeatedly to improve the wiring.

The EDA placement and routing tools are not as smart as an experienced human designer. With literally millions of transistors on a chip, however, the tools do most of the layout. Nevertheless, sometimes we must intervene manually to complete a layout.

Nora: Okay, so the placement affects the routing and the routing influences the placement? How do you know when you are done?

Larry: Basically, when all the wires are routed and no critical signal wire is too long. (And all the other complex factors pass checking.)

Placement and routing tools try simultaneously to minimize several factors. These include the area, wire lengths, the number of wiring layers, and *vias (*layer-to-layer connections). Here, let me sketch this on a napkin for you. (See Figure 1.4.)

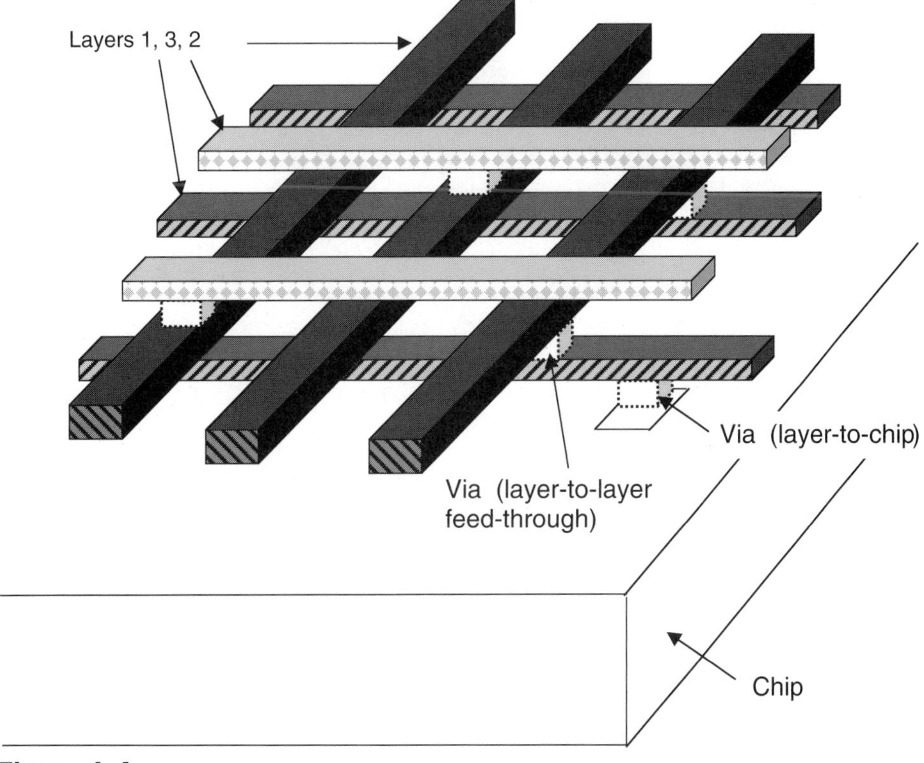

Figure 1.4
IC Wiring

Larry: See, the IC is built in layers. I didn't show the transistors, but the transistors are on the bottom layers on the chip, and the wiring is above. I drew only three metal wiring layers, but there might be four or six or more. Insulating material separates the wire layers. The only connection between wires and the transistors are little vertical metal posts called *vias*. You can see that electrons traveling between transistors may have to go along several wires on different layers, and through several vias. That is the routing problem.

After successful place and route, I turn the design back to Andrea, the ASIC engineer. She runs all kinds of analyses, such as timing, performance, power, area, thermal, and so forth. Then she runs all kinds of manufacturing design rule checks. These include things like sizes of transistors, wires, spacings, vias, and so forth.

Did You Know?

Correcting an IC error after manufacturing is very expensive and time-consuming. It may cost millions of dollars and take up to a year. Companies have gone out of business because of deadlines missed due to a chip error.

Designers need to be sure *everything* is right before going to IC manufacturing. EDA tools warn them of errors (such as a missing connection) and power, timing, or thermal problems.

Nora: I didn't realize the errors were so costly! No wonder there are so many checking steps.

Larry: Yes, and we have more checks required every time we improve the manufacturing process and things get smaller. New kinds of problems show up and the EDA tools have to adapt and check for them. New EDA tools are being developed all the time to keep up.

Nora: What happens after all the checks are done?

Larry: All the placement and routing information overlays the base chip. The base chip has all the bottom layer transistor information, and the input/output (I/O) pads. The final physical operations involve *merging* the design layers onto the base chip layers.

Then we create a final design database with all the information on it. A mask-making shop uses the final database to create the masks for the IC manufacturer.

This database file used to be stored on a magnetic tape, and the tape was sent to the IC manufacturer. The step of creating a final clean design database is still referred to as *tapeout,* and is still a really BIG event. The company may announce the new product and often celebrates with a party.

Nora: So you are all involved in taking a system product from idea to an integrated circuit chip? And there are EDA tools at just about every step?

Andrea: That's right.

Nora: And I suppose there's another whole series of steps to actually manufacture the integrated circuit?

Andrea: Right again. And another party when the first manufactured chip (first silicon) works.

Nora: Okay, I will save those questions for another party. I have to digest what I heard here before I can learn any more.

Sanjay: (Nora's supervisor) You asked yesterday about the benefits of EDA. Do you know some now?

EDA BENEFITS

Nora: Well, EDA helps a lot of people. It helps the system designer use the latest technology and explore different design approaches. EDA programs allow the designer to model the system's performance and estimate its power needs.

It helps the ASIC logic designers put their logical design ideas into computer form. It also helps them handle the high complexity of integrated circuits.

EDA helps the layout designer place and route millions of transistors on the IC. It helps check hundreds of physical and electrical design rules for all those transistors.

EDA helps product companies achieve more complex chips with lower cost, and shorter time to market.

It helps all of us, wherever we have electronic systems—like in cell phone and satellite communications, TV and personal entertainment products, smarter cars, and even in military systems.

SUMMARY

ICs are small devices that make electronic products work. They contain millions of transistor devices and connecting wires.

EDA tools are computer programs that help chip designers do their jobs. For example, they help designers specify devices and simulate how they will operate together. These tools are like programs that architects use to design buildings or that civil engineers use to design bridges.

The designers could not handle the IC complexity without the tools. Moreover, the number of transistors per IC continues to double every two years or so. The smaller dimensions cause more complex electrical issues that EDA tools must check for or avoid.

Creating a chip involves many stages, including specification of what the chip will do. There are different EDA tools for each design step and task in the design sequence.

The logic design stage determines how the chip outputs will depend on its inputs. Simulation and modeling test the chip behavior with a computer model. Synthesis translates the logic design into actual devices. Layout places (locates) the devices on the chip and all the interconnecting wires.

A chip design team typically consists of many engineers. Some work on a single stage of design, others work on several. All of them need EDA tools to do their jobs.

An error in the final tapeout can be very expensive in time and money to fix. A single tiny error can fail the entire chip, so exhaustive testing is required.

There is more detail on electricity, semiconductors, and logic in Appendices A, B, and C. Appendix F supplies additional background on the semiconductor and product design industries.

2 The Business of EDA

In this chapter...

- Introduction
- EDA User Return on Investment
- EDA Vendor Return on Investment
- EDA Tool Development Sources
- The Time-to-Market Competition
- EDA Business Models
- EDA Industry Growth
- EDA People and Conferences
- Summary
- Quick Quiz

INTRODUCTION

This chapter describes the business side of EDA. We explore the financial and market features, and where and how EDA tools are developed.

The business models are significant for understanding the EDA customers and how EDA companies make money. A healthy EDA industry is essential to the electronics industry, and vice versa.

The developers try to anticipate the designer's needs, but typically they are two years behind. This is not too surprising since the chip complexity and density continue to increase rapidly. Imagine the problems that architects would have if building sizes doubled every few years. In addition, the new chip technology brings new problems to light. It takes the IC engineers a long time to identify the problems needing EDA help.

Leading chip designers lobby for certain features. If enough customers ask for the same feature, developers may eventually add it to the tool. Perhaps a new tool may be created.

Let us listen in as Nora talks to Frank, the chief financial officer (CFO) of Sandbox, Inc.

Nora: Thanks for taking the time to meet with me. I am trying to understand the business side of EDA. Sanjay, my supervisor, suggested I talk with you. I talked to several engineers at the party yesterday and got some understanding of the EDA tools.

Frank: So, what would you like to know?

Nora: Well, to start with, how big is the EDA industry?

Frank: It is much smaller than the traditional software industry. EDA's size is a small percentage of that of the semiconductor industry that we serve.

Did You Know?

In 2002, the worldwide EDA industry had a total revenue of about $4 billion dollars. It employed about 18,000 people. By comparison, the semiconductor industry revenue was about $140 billion, employing 270,000 people.

Nora: How many EDA companies are there?

Frank: There are about 200 companies and about 50 different kinds of tools. Interestingly, most are very small, like Sandbox. There are only a few large EDA companies with revenues over $500M and 3,000–5,000 employees.

Nora: Why are there so many EDA companies, if it is such a small industry?

Frank: The continuing semiconductor IC evolution drives the IC designers to need more EDA tools. This is because the number of transistors per chip has doubled about every 18 months. This has led to more capability and lower cost on a chip at the same time.

Did You Know?

The transistor density doubling every 18 months was observed by Intel's Gordon Moore in the 1960s and has continued for decades. This statement is commonly referred to as *Moore's Law*, and it represents the incredible semiconductor technology development.

Nora: So the ICs get steadily more complex?

Frank: Yes, and that makes them harder to design. There are more functions on the chip, and the transistors number in the millions. This is where the EDA tools come in.

Nora: How do they help?

Frank: Well, the tools make the designers much more productive. ICs used to take two or more years to design. Now they are much more complex, yet the designers are asked to complete them in a few months. There is more capacity on the chips than the engineers can design in the needed time. Their productivity is measured by how much they can design over time. This has become a real issue in recent years. Let me draw you a quick sketch. (See Figure 2.1.)

Nora: So the designers are not keeping up with the semiconductor technology?

Frank: That's right. There has always been a gap, but it appears to be increasing. Part of the problem is that the needed time-to-market (TTM) has been shrinking. The only help the IC engineers have are the EDA tools. That is why the IC designers are always driving EDA development. There are many EDA companies because most are small start-ups responding to this demand for better and faster tools.

Nora: Why are the EDA tools so important to our customers?

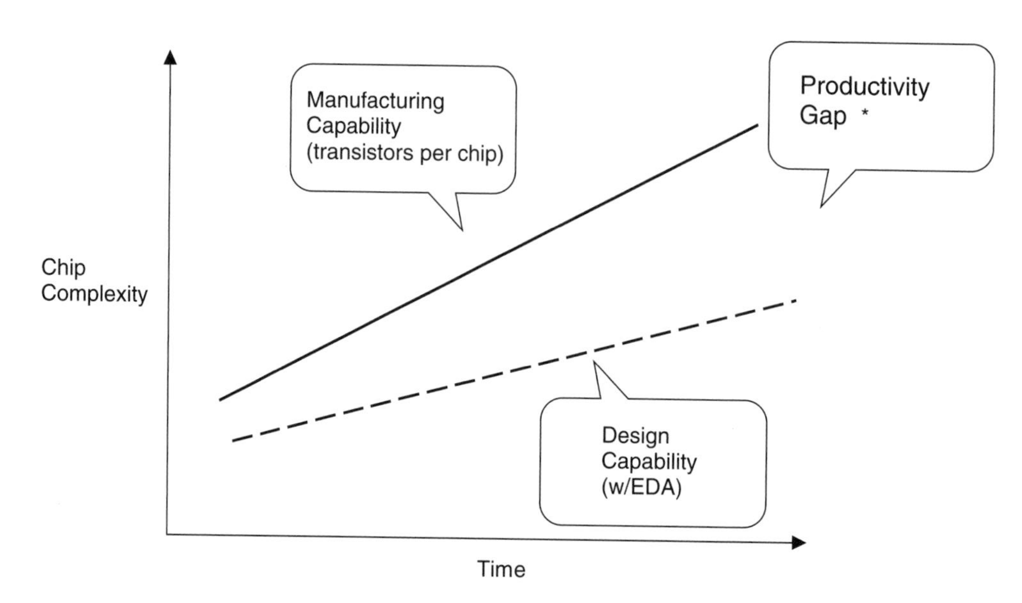

Figure 2.1
Productivity Gap

EDA USER RETURN ON INVESTMENT

Frank: Well, suppose an EDA tool can reduce product design time by a week or a month. The electronic product company can gain millions of dollars. This is due to four factors (in order of importance).

First, there will be a longer product life due to earlier TTM. Many electronic products generate millions of dollars per month during their lifespan. So any increase in the market window is significant.

Second, product manufacturers can get a much higher price for early products, as in most industries. With little or no competition, the vendor can set the price.

Third, the product may become the market leader simply by being early. The market leader gets the major share of the market.

Fourth, there is reduced product development cost. The shorter time the engineering team spends on the product, the less cost there is to recover.

EDA tools thus have a potentially huge *return on investment (ROI)* for the Original Equipment Manufacturer (OEM) electronics product companies. This is why semiconductor and OEM companies spend millions of dollars on them. Investing in EDA development can be a tremendous benefit. (But it can also be a large sinkhole into which you throw money!)

Nora: Can you give me an example?

Frank: Sure. I once did a quick estimate of EDA research value for a major semiconductor company. We were trying to reduce development time by several months. I looked at the value to them if a product got to market one month earlier.

Usually they would not design a chip unless the revenue expectation was about $1M/month. Since the price is higher in the beginning, extending the product life by early market entry is best. Let me summarize the assumptions and results on the whiteboard. (See Figure 2.2.)

EDA User ROI Example:

Assumptions:

Revenue Goal: $1M/month IC (about 100,000 ICs at $10 each).

Production Life: Useful life continues for an average of 24 months.

Number of IC designs: One $10 chip

Results:

Value of one additional month of sales life for ***one*** IC design = $1M!

Value of one additional hour of sales life = $6,250! ($1M/160 hr./mo.)

(Not a bad return for someone earning about $60/hr.!)

Value of one EDA tool for ***ten*** designs = $10M!

Value of the EDA tool to the user?

$50,000 - $100,000 - $500,000!

Figure 2.2
Example—EDA User Return on Investment (ROI)

Now, not every customer is a large semiconductor company. Electronic product companies may have only 1,000 sales per month. However, their product price is much higher (say, $100). So their potential ROI is only (!) $100,000 per additional month of sales life.

Nora: So you are saying that EDA tools can create big revenue gains by increasing the life of electronic products and getting to market earlier.

Frank: Yes, that's the biggest of the ROI gains for the EDA user.

Nora: What is the ROI for us, Sandbox, as an EDA vendor?

EDA VENDOR RETURN ON INVESTMENT ....

Frank: Let's look at that. Let me draw on the whiteboard here. (See Figure 2.3.)

EDA Vendor ROI Example:

Assumptions:

Staff: Five software engineers

Development time: Ten months for a new EDA software tool

Average salary and benefits cost per employee: $10K/month

Total available ASIC designer market: 5,000 users

Expected market share: 15% (750 copies total)

Product life: Two years

Company size: 35 people (development, administration, etc.)

Results:

Total development cost: $500,000 (5 x 10 x $10K)

Revenue needs: $500K development +

$7.2M company support (30 x $10K x 24 months)

= $7.7M

Break-even cost per copy: $10,267 ($7.7M/750)

Break-even cost with overhead (facilities, utilities, etc.): ~ $20K

Figure 2.3
Example—EDA Vendor Return on Investment (ROI)

Now this is admittedly a rough estimate, but the tool prices come out about right. We can typically sell a tool for anywhere from $10K to $200K per copy, depending on how critical it is. So you can see there are potentially large profit margins in this industry.

Nora: I see a large difference between the development cost and the value to the user. You can sell a single tool for ten times what is needed to support the company?

Frank: Sometimes, but it depends on the perceived value of the tool and the amount of competition. Only a few tools sell in the multi-$100K class. Also, larger user companies often have a larger budget for EDA tools than small design houses have.

However, an early time-to-market product can demand a higher price. In addition, an aggressive marketing and sales staff can increase the number of copies sold.

EDA TOOL DEVELOPMENT SOURCES....................

Nora: So who develops EDA tools?

Frank: Some tools start as a university student's Master's or PhD thesis. Many of the ideas come from the professor, but the students do most of the work. A few semesters of thesis work is enough to prove the feasibility (or not) of an idea. It may take contributions by several students over a few years. However, the actual university software code is rarely ready for *production use*.

Later, the graduate or others may develop the software into a *production tool*. (A production tool is well-documented, exhaustively tested, user-friendly, stable, and benchmarked against other tools.)

Did You Know?

Many companies fund university research. It takes only a $20K–$30K contribution to a professor in most schools. That gives the company open access to learning about all the research going on. Most schools have industrial liaison programs with periodic presentations about their research projects.

EDA research support also comes from government sources. In the U.S., these include the Advanced Research Projects Agency (ARPA) and the National Institute of Science and Technology (NIST).

Small EDA startup companies develop some tools. These companies may be a *spin-off* from a university or from a large EDA company. They may come from the internal EDA development group of a system or semiconductor company. Other sources of EDA tools include consultants, design services companies, and government laboratories.

Large EDA companies acquire EDA tools through internal development or as extensions of existing company EDA products. They also get tools from universities or by merger with or acquisition of another EDA company.

Many small EDA companies develop specialty tools for secondary IC problems with modest initial markets. As ICs shrink, the secondary effects become vital. A larger EDA company often then acquires the tool or the company.

In-house/Out-source EDA Tool Development

Nora: Why don't our customers develop their own tools?

Frank: Some do. They may develop the tools for internal use and then find there is outside interest. They can sell or lease the tool. Alternatively, they may spin off a new company to exploit the market (and get royalties or stock). Sometimes the EDA team may leave and form a company to productize the tool.

However, internal EDA development groups have a lot to do. They support existing tools, test tool upgrades, solve bug problems, and service many groups. As usual with service groups serving multiple masters, they have to prioritize their time. So there is little time left for actual new tool development. In addition, small companies have a more difficult time supporting an EDA group.

So many user companies outsource their EDA development to independent *third-party companies*, such as Sandbox. The third-party vendors try to amortize the tool support and service cost over many customers.

At first glance, those companies specializing in EDA should be able to do it better. They get input from many sources and can spread the cost over multiple users. However, EDA vendors have the same priority problem as the users' internal EDA groups.

Many users use EDA vendors to reduce the size of their internal EDA support staff. For others, the vendor response is too slow. For critical projects, management can focus internal resources but cannot control the vendor priorities. Some companies (computer makers in particular) need specialized EDA tools not available from EDA vendors.

So a few large companies (e.g., Fujitsu, IBM, Intel, Motorola, National Semiconductor, NEC, and Texas Instruments) still have internal EDA development efforts.

Nora: Okay, I see there are many EDA tool sources. I gather there are different kinds of ICs. Do they all use the same tools?

Frank: Not all. Although many tools are common, different IC *architectures* require different suites of tools. Let me explain the different architectures briefly, since that is another view of the EDA business.

THE TIME-TO-MARKET COMPETITION

Frank: In full custom design, all the IC parts are optimized to give the fastest and densest layout. Early ICs were all small custom-designed parts. They were sold as standard building blocks to multiple buyers. The buyers would use them to build many different kinds of products.

Since time-to-market is so important to ICs, companies invented *semi-custom architecture* to speed up chip design. The IC *architecture* is how the transistors are structured (or arranged) on the chip. The idea is to make most of the chip standard, with some *tailoring* (customizing) done for each customer application.

An analogy is furniture with a common design, but a wide choice of cover material. Another analogy is a new car with a standard basic design but with many colors and options. If you buy what they offer on the lot, you can get it quickly.

There are three main kinds of semi-custom architecture: gate *array (GA), standard cell (SC),* and *field programmable gate arrays (FPGAs).* Each requires some specialized EDA tools. Either the manufacturer or the customer customizes each kind.

The GAs and SCs are called Application Specific Integrated Circuits (ASICs). (FPGAs are not ASICs per se, but compete directly with them.) Each ASIC is customized for a specific customer.

***Gate Arrays*:** The IC is covered with an array (fixed rows and columns) of identical logic gate circuits defined by the manufacturer. (Think of rows of identical cars in a rectangular parking lot.) The customer defines how the gates should be wired together. It is the wiring, implemented by the manufacturer, which customizes the gate array.

***Standard Cells*:** The customer chooses logic *cells* (groups of gates) from a *library* of standardized cells. A cell can be much larger and more complex than a simple gate. (Think of trucks in the parking lot instead of cars and choosing the type of truck for each parking spot location.) The customer defines how they should be wired together. Thus both the choice of cells and the wiring are custom. The manufacturer implements the cells in the chosen locations on the chip and does the wiring.

***Field Programmable Gate Arrays (FPGAs)*:** These are fixed arrays of complex gates defined by the manufacturer. The interconnect wiring of these gates is a fixed multilevel matrix or mesh of wire segments defined by the manufacturer. The customer defines the function of each group of gates or logic cells, and how they should be wired together.

Each possible connection point is a selectable fusible link or a switch controlled by a memory cell embedded in the chip. The customer defines how they should be wired together. EDA software then chooses the right combination of wire segments and connects them up by selecting the needed link or switch points.

The same FPGA chip is mass-produced for many customers. Each customer then does the custom wiring in seconds at their own facility or in the field. This makes for very a fast TTM.

The fusible link programmable chips are usually programmable just one time. The memory-based switch type chips can be reprogrammed for easy design modification. Most FPGA memory-based chips lose their memory when power is off (*volatile*), but some use *flash* memory to be *non-volatile*.

However, for the same function, FPGAs may take 10-20X more chip area than SC ASICs. Therefore, they are more expensive and slower for the same amount of circuitry. Vendors make many variants of each architecture.

As I mentioned, each of these architectures uses different EDA tools, particularly for the physical layout design. The FPGA vendors usually supply their own tools, free. Each company's tools are optimized for its architecture. Only a few FPGA tools are from pure EDA companies.

Nora: How do the engineers decide which approach to use?

Frank: Customers choose an architecture depending on their IC complexity, TTM, cost, and volume needs. The cost/volume crossover point depends on the application, of course. But a recent rule-of-thumb had FPGAs as more cost-effective up to about 1,000 units.

If a change is required, the ASICs need to rework (*re-spin*) the silicon, whereas the FPGAs do not. SCs may require all the mask layers to be redone, and GAs require only the top metal interconnection layer (partial re-spin).

Each approach has its merits and the competition is fierce. Heated debates occur on conference panels about which is the better approach. This competition is part of the business landscape for EDA tools.

Let me sketch you a quick table for comparison. (See Table 2.1.)

Table 2.1 ASIC Architectures

STYLE	Standard Cell	Gate Array	FPGA
Relative Area	1X	2X	10X
Relative Speed	Fast	Medium	Slow
Gates Available	High	Medium	Medium
Low-Volume Cost	Very high	High	Low
High-Volume Cost	Very low	Low	High
Mask Cost	~ $500K+	~ $300K+	Not customer cost
Design Time	3-6 mo.	3-6 mo.	< 1 mo.
Tool Cost	High	High	Low to free
Ease of Change	Full silicon re-spin ~2 mo.	Partial silicon re-spin ~ 1 mo.	No silicon re-spin < 1 day

Note that for low-volume and low-speed applications, the FPGA's low cost and short TTM win. FPGAs also fit applications which are likely to change frequently. Military and aerospace firms are major FPGA users. There are also EDA tools to convert an FPGA to an ASIC, should the demand and number of units needed increase.

Note that the SC usually wins for high volume, stable, medium performance applications. (GA use has steadily declined.)

Custom chips are still used for very high-volume, high-performance, or very low-power applications.

Nora: The IC variations are more complicated than I realized.

Frank: Well, IC types develop over time. I mention ASICs because most ICs are usually one type of ASIC or another.

There are also combinations with FPGA blocks on an SC chip. Some chips have SC blocks embedded in an FPGA. This can be a problem since some design steps and EDA tools differ for each architecture. And other chip architectures embed increasing numbers of memory blocks into ASICs and FPGAs.

EDA BUSINESS MODELS ....

Nora: Do the business models for the EDA vendors vary as much?

Frank: Well, they vary, but not as much.

Developing the next leading-edge product while supporting your current products is a high-risk venture.

Did You Know?

The leading product company in one technology generation will usually *NOT* be the leader in the next generation. That is an interesting observation that applies for all kinds of businesses, not just EDA.

Many large and small companies compete, but at any given time only one of them has the best-in-class product. Small companies are more likely to succeed since they can focus their resources on one product. (There is no cushion—they must sell the new-generation product or go out of business.)

Moreover, vendors do not want to lose the revenue from existing products. In addition, customers are often slow to migrate and want continued support and upgrades. A large vendor has more support staff for existing products. That may limit resources available for risky (but essential) new product development.

Nora: How easy is it to market a new tool?

New EDA Tools

Frank: It can be difficult for a small developer to sell its EDA tool. Small customers prefer to wait until a large respected user has tried a new tool. Most users want a reference or some success history. With such a reference, an EDA tool vendor has a much better chance to sell a new tool.

However, getting a large customer to try a new tool is also not easy. The customer design engineers (and management) are usually very risk-adverse, since they are paid for on-time success. They have a deadline to meet. They know how to use their present tools. They don't know how well (or if!) the new tool will work.

There is also a substantial learning curve to use a new tool. It can take several months to learn a new EDA tool. The engineers have to try it out and compare it to other new or existing tools. (Internal development groups face the same hurdle, by the way.) The company has to invest time, money, and staff.

Therefore, an EDA tool vendor must show that their tool gives a large (maybe 5X-10X) improvement. This is usually in *run time*, performance, design size, or accuracy. Sometimes customers will try a tool simply because there were no other options available to meet their project deadline.

At the same time, there is a symbiotic relationship between the EDA vendors and the users. The large users need the latest tools for their innovative designs. Most customers need new tools frequently to keep up with increased chip complexity, speed, and new issues. The vendors need the users to check out the tool and encourage other buyers.

Nora: Well, marketing a new tool sounds like an interesting challenge. How will Sandbox grow?

Licensing Models

Frank: A great deal of company revenue depends on the licensing model it uses for revenue. The company can sell the tool, license it for a fixed or variable time, license it to a customer company or individual user by calendar time or usage, etc. The choice of a competitive and profitable license model can make or break the company. Licenses are often negotiated on a per company basis.

Yearly maintenance charges (often 15% of the sales price) for updates and bug fixes also contribute significantly to the revenue and cash flow picture.

Mergers and Acquisitions

Frank: In order to grow, a small tool developer must either be acquired or merge with other small vendors. Acquisition is usually by one of the three or four large EDA tool vendors.

Alternatively, the small vendor needs to develop a series of successful new tools. Historically, as I said, the chances of the same group developing another "star" tool are very low.

For the large tool vendor or OEM, the problem is similar. It is hard to maintain a leadership role. Acquisition of small promising startups is one way that companies catch the "next wave."

However, the integration of one company into another is always tricky. Different cultures, expectations, and management styles make integration from mergers and acquisitions difficult. There is often the issue of a choice between competing internal products and acquired external products.

Many promising EDA tools (and their developers) disappear after an acquisition or merger. There are some ways to sweeten a merger. However, there is no way to ensure that key technical people will not leave.

An example of such a merger is when Synopsys, a large EDA company, acquired Avant!. The Synopsys culture runs on consensus, encouraging employees to speak out. At Avant!, employees reportedly were expected to do as they were told.

In addition, some Avant! tools were complementary to Synopsys' tools and others were competitive. There were some hard choices on which tools to keep and which to drop.

Another example was when Valid Logic merged with Cadence. Cadence attempted to alternate management layers. For cultural reasons, this did not work very well. Groups often form a strong sense of identity, which can conflict with such a shakeup.

Nora: Perhaps "explosive growth" doesn't apply to the EDA business. Are there other business models?

Application Service Provider Model

Frank: Yes, a few others come to mind. The Internet has led to another business model—that of the *Application Service Provider (ASP).* In this model, the EDA vendor hosts a tool or service on the Internet. The vendor sells the tool or service on a usage basis. (Either by time or some program completion event.)

The user does not get the tool source code or use of the tool on their own machine. The ASP vendor provides a server on the Internet.

Nora: I imagine that should be a major growth area?

Frank: Possibly—although remember that most EDA involves very large, secure data files and transfers. The Internet is not so good for those.

Design Services Business

Some EDA companies also try to expand their business models with Design Services divisions. They provide designers or consultants to help an electronic product company execute a design. They will help with or do the whole design. Design services can complement the EDA tool side of the business, providing a testing ground and reference.

Design services, however, do not have the same *leverage* as software EDA products. A small software staff can create a product that generates a lot of revenue. The revenue is not in direct proportion to the cost. That's leverage—creating a lot from a little. To grow in design services, you have to increase the staff in direct proportion to the revenue. If sales fall off, you still have a large staff to feed.

Other EDA companies have acquired and sell re-usable design blocks, called *Intellectual Property* (*IP*). Re-use of predesigned blocks saves a lot of design time as chips get more complex. Therefore, this practice has been growing slowly but steadily. The IP business has too many pros and cons to go into here. (The reader can learn more about IP in Appendix F.)

Some companies form or join standard groups trying to promote their tool approach as an industry standard. Others form alliances with vendors of complementary tools to offer a more complete tool suite for their customers.

Nora: Thanks, Frank. There really are quite a few ways to make money in this business.

Frank: Yes, but remember the EDA business is very small compared to its customers.

Nora: How do they compare?

EDA INDUSTRY GROWTH

Relative Industry Sizes: EDA, IC, Electronics

Frank: EDA is a critical part of the whole electronic product food chain. One might think it should represent a larger part of the electronics revenue. However, it is a very small part revenue-wise.

This question does indeed come up almost every year at EDA conferences. They often have a panel discussing why the EDA industry is not growing faster and larger. They ask why it does not receive a larger percentage of revenue from the semiconductor industry.

This is a little like comparing the price of a saw or screwdriver with the price of the house. The tools are certainly critical to the rapid building of the house and thus add great value. However, they are a small piece of the entire housing industry *food chain*.

The house food chain includes the costs of labor, material, land, foundation, and plumbing. There are contractors for electrical, frame, roof, walls, and painting. There are costs for permits, inspections, subcontractors, sales, and finally, the tools.

Like the house analogy, a long food chain contributes to the electronic product price. The chain includes product design, package design, product manufacturing, testing, and marketing. There are costs for sales, support, distribution, and inventory.

Further down the chain are the integrated circuits. They are one (small) part of the product manufacturing. The IC food chain includes chip design, verification, manufacture, assembly, and test. Add to that marketing, sales, support, distribution, and the EDA tools.

Thus, there is not a lot left over for the EDA tools. The EDA market is only a few percent of the semiconductor market. That market is itself only a small part of the ~ $1 trillion electronic products business. If I draw a comparison chart, you will see how small that is. (See Figure 2.4.)

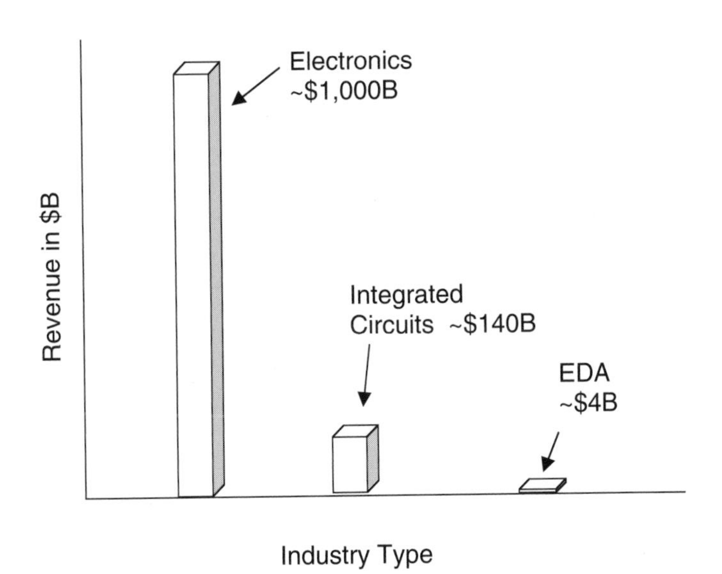

Figure 2.4
Relative Industry Revenues

Nora: Yes, I see, it is tiny compared to the IC and the electronic product revenue. Aren't the EDA folks taking a lot of risk developing the tools?

Relative Risk Factor

Frank: Actually, they are not. Risk is a factor in the relationship of EDA tools to the overall product food chain. However, the company selling the end-user product has most of the risk and product costs.

The semiconductor company owns most of the risk and capital outlay developing the product ICs. (It builds wafer fabs which cost billions and take years to build. Then, the market may not need them.) In developing software tools, the EDA tool vendor incurs relatively little risk or cost.

Nora: Thanks very much, Frank. I think I have a much better feel for the EDA industry now.

Frank: Glad to help, Nora. Good luck in your new job.

EDA PEOPLE AND CONFERENCES

Nora: (back at her desk) Hi, Sanjay. I just finished talking with Frank, the CFO. Thanks for the suggestion. Are there any other key industry players I should be aware of?

Sanjay: Yes, plenty of them. Although small, the EDA industry has had some colorful personalities in its brief history. Many interesting people have started or worked for EDA companies. Some have been brilliant researchers, gifted marketeers, and others have not been considered "paragons of virtue"! Some have started companies, and some have sold their company and gone off to start or lead another. The migration of people within the EDA industry is an interesting story.

A while ago, there was a very colorful lawsuit in the news. Here is a summary of that.

THE AVANT! DEBACLE

Gerald Hsu worked as a vice-president at Cadence with a goal of driving a small tool competitor out of business. He ended up joining the competitor as CEO, and he formed a new company—Avant! Avant! grew rapidly to provide a fairly complete tool set using a central common database (something their competitors lacked).

However, one of Hsu's new employees came from Cadence, allegedly with the software for a key Cadence tool. According to the prosecutor, the case was unique in that there was a large, publicly traded company that was founded and built on stolen property.

A long, ugly lawsuit evolved. Cadence won, millions in fines were levied, and people went to jail. Four top Avant! executives went to prison, but not Hsu.

The case dragged on for six years. The defense team got three judges to excuse themselves. Avant! ran an extensive media campaign to promote its side.

Avant! was eventually bought by Synopsys, Cadence's arch-rival. The episode illustrates both the value of trade secrets and the difficulty of keeping them secure, even internally.

Many EDA companies are constantly involved in legal cases. These are typically related to patents, trade secrets, sales agreements, or partnerships. Legal cases can be draining on resources and executive time. Cases often drag on for years. The Avant! case set a precedent within the software industry for successfully prosecuting the theft of source code.

Nora: You are right. EDA is more colorful than it first appeared.

Sanjay: Other people honored for their contributions to EDA have been awarded the Kaufman Award. The Electronic Design Automation Consortium (EDAC) sponsors this award.

People Opportunities

Nora: I noticed that the industry seems to have CEOs from every part of the world.

Sanjay: You are very observant. As an industry, EDA is exceptionally open to minorities and women. Most large EDA companies are based in the United States. However, there are some European and Asian companies. Many companies have research, design, and support groups in Europe, Israel, India, Japan, Singapore, China, and Taiwan.

Advancement is mostly based on merit. There are few "glass ceilings" or "old-boy" networks compared to other industries. There is also a natural liaison with academia for both staff and startup ideas.

As a startup business, little capital equipment is needed. Venture capital is primarily used for staff costs. Without plant or manufacturing costs, developing EDA software programs can earn a good return.

Nora: That's encouraging.

Key Conferences

Nora: Are there any EDA conferences that we participate in?

Sanjay: Yes, several. Key EDA conferences include:

Design Automation Conference (DAC). The largest EDA conference has an unusual blend of industry and academia. There are technical sessions, panels, elaborate exhibits with professional entertainers, and large parties. There are canned demonstrations (demos) and serious customer private sessions in closed cubicles. It is held in June in different cities.

International Conference on Computer Aided Design (ICCAD). This conference features mostly academic and technical papers. It is held in San Jose in November.

Design Automation and Test Europe (DATE). This is a major EDA/Test conference. It offers a mix similar to DAC, but much smaller. It is held in Europe in February.

International Test Conference (ITC). The major test and test equipment conference. It is held in September or October, usually in the United States.

(See Appendix E for more detail on conferences.)

SUMMARY

EDA is a relatively small industry compared to the electronics, semiconductor, and software industries. However, it is a critical part of the success of these industries.

There are some 200 companies and over 50 kinds of EDA tools. EDA companies must continue to innovate to keep up with the on-going and rapid advances in the semiconductor industry.

EDA users are willing to pay for expensive EDA tools to achieve the TTM and design complexity which their products require.

EDA companies can make a good ROI because a relatively small work force investment can result in sales of an expensive, needed product.

EDA companies usually start out as spin-offs from existing companies or as startups from university research.

Some product or semiconductor companies develop their own tools, but a majority use EDA vendor tools.

The competition and TTM needs have resulted in several different IC architectures. These include Gate Arrays, Standard Cells, and Field Programmable architectures. Each (and their variations and mixtures) fills a need, depending on the application.

There are a variety of EDA business model issues such as new tool development, licensing models, mergers and acquisitions, application service provider, and design services.

EDA is a small but essential piece in the growth of the semiconductor and electronics product industries.

QUICK QUIZ

1. What is the main source of EDA ROI for the EDA user?
   a. License fees
   b. Increased product market life
   c. Sale of IP

2. What is the main ROI for the EDA vendor?
   a. High-priced tool sales
   b. Maintenance fees
   c. Venture capital

3. Which is NOT a source of EDA tools for the user?
   a. University tool
   b. In-house development
   c. Vendor outsource
   d. Microsoft

4. Which is better for low-volume, fast TTM design?
   a. Full custom
   b. Gate array
   c. Standard cell
   d. FPGA

5. Why will a large EDA company buy a smaller EDA company?
   a. To add new tools to its line
   b. Tax write-off
   c. To lower its IC fab costs
   d. To shorten its IC design time

6. How large was the EDA business in 2002?
   a. $6B
   b. $4B
   c. $100M
   d. $300B

7. Which company takes the least risk?
    a. Electronic product company
    b. Semiconductor company
    c. EDA company

8. What did the Avant! case involve?
    a. Computer theft
    b. IC theft
    c. Accounting cover-ups
    d. Trade secret theft

9. What is the largest EDA conference called?
    a. ICCAD
    b. ITC
    c. DAC
    d. DATE

Answers: 1-b; 2-a; 3-d; 4-d; 5-a; 6-b, 7-c; 8-d; 9-c.

3 The User Perspective

In this chapter...

INTRODUCTION

This chapter gives the EDA tool users' perspective on the tools and the industry. Who are the users? They are the electronic systems, semiconductor, or ASIC development engineers who design ICs. They are also the engineering design and EDA administration and support managers.

New EDA tool customers need to be aware of the significant costs and risks involved. We will consider tools, tool integration, computer choices, and staff.

Since the IC technology shrinks so rapidly, EDA tools are continually updated or replaced to handle more complexity and new design issues. Tools are normally rushed to market with increased capacity for more complex chips, speedups, and new capabilities.

Users need to expect some frustration in evaluating tools, tool performance, vendor support, in-house support, and documentation. As one might expect, users often find problems which the tool designers never imagined. Everyone is learning at the same time.

Nora's supervisor Sanjay has sent her to Hugo, an EDA tool user manager at Fabless Design, Inc. Nora will be working on his account for Sandbox. Let's listen in as Hugo shares his experience with Nora.

Nora: Hugo, thanks for taking the time to explain what EDA users need. What kind of problems do you deal with?

FOUR KEY EDA USER DECISIONS

Hugo: Nora, we first need to clarify some issues before we talk about tools. An IC design manager or the EDA manager has to deal with four main questions. (I act in both roles here, by the way, as a design manager and as an EDA manager.) Those questions are:

1. What is the design organization (who and where)?
2. What kind of network capability will you need?
3. What are the security requirements you must meet?
4. What kind of computer systems do you need?

I listed them in this order because each depends somewhat on the previous answers. The tools used are affected by the answers to these questions. Let's take them one by one.

Organization

1. First, look at the organization. Will one or two people or a group do the design work? Will they all be at the same location or in different remote places? For example, we have a design group in Ireland.

   Running design jobs remotely is common on private networks or on the public Internet. This has been an area of tool development for many years. (Some EDA companies provide network support utilities.) *Distributed design* allows you to partition or share the work between design groups separated by time or geography.

   Will you do the design work at different sites on the same "campus" (metro)? Perhaps you will do it in different time zones or at different geographic sites (same or different countries)?

Hugo: (chuckling) I remember trying to set up a video-conference time common to Japan, the US, and the UK. Someone was always inconvenienced.

Did You Know?

Coordinating design groups working around a country or around the world is often very difficult. Just scheduling a time for a video conference or a conference call is complicated. For instance:
At 9 A.M. in San Jose, people are just starting work.
It is 5 P.M. in London; people are ready to leave work.
It is 2 A.M. the next day in Tokyo!

However, the massive size of many EDA files is a deterrent to the rapid transfer between remote sites. Since EDA files contain proprietary chip designs, protecting them is a major concern, particularly over the Internet. For example, we have multi-gigabyte files that can take hours to send or even longer, if there is a network failure and we have to resend them.

File sharing also requires identical or compatible EDA tools at the local and the remote site. So you also need consistent design, change control, and backup procedures.

However, the biggest challenge is the cultural and personnel issues, even within the same company. For example, I once had a design group in Japan working with one in the US. Just following the same checking procedures and agreeing on deadlines was a real problem.

Will the various users doing separate parts of the design be using different software or computer systems? (They may have to be the same, and the same revision.)

If you can answer these questions, then you can determine the sort of computer network you need.

Computer Network

2. Second, look at the structure of the computer network. What does the organization already have, and what does the engineering staff need? Most design engineers like to run or control the tools from a workstation at their desk. (This is important for interactive work. For simulation runs, the tool will run faster on a shared *server* computer or a group of shared computers (*server farm*) with massive amounts of memory.)

   If there is a design group, you may want to share a local engineering server. Is a company-wide shared server on-site? Are you required to interface with it? Do you need a *local area network*? Is there one in place, and is it fast enough?

   Is there a geographically remote server, involving high- (or not so high-) speed communications lines? Is a connection to the Internet required for remote communications? (Some countries do not allow their residents open access to the Internet...)

   If you can define the computer network you have or need, then you can address the security questions. Let me draw you a quick sketch of computer networks. (See Figure 3.1.)

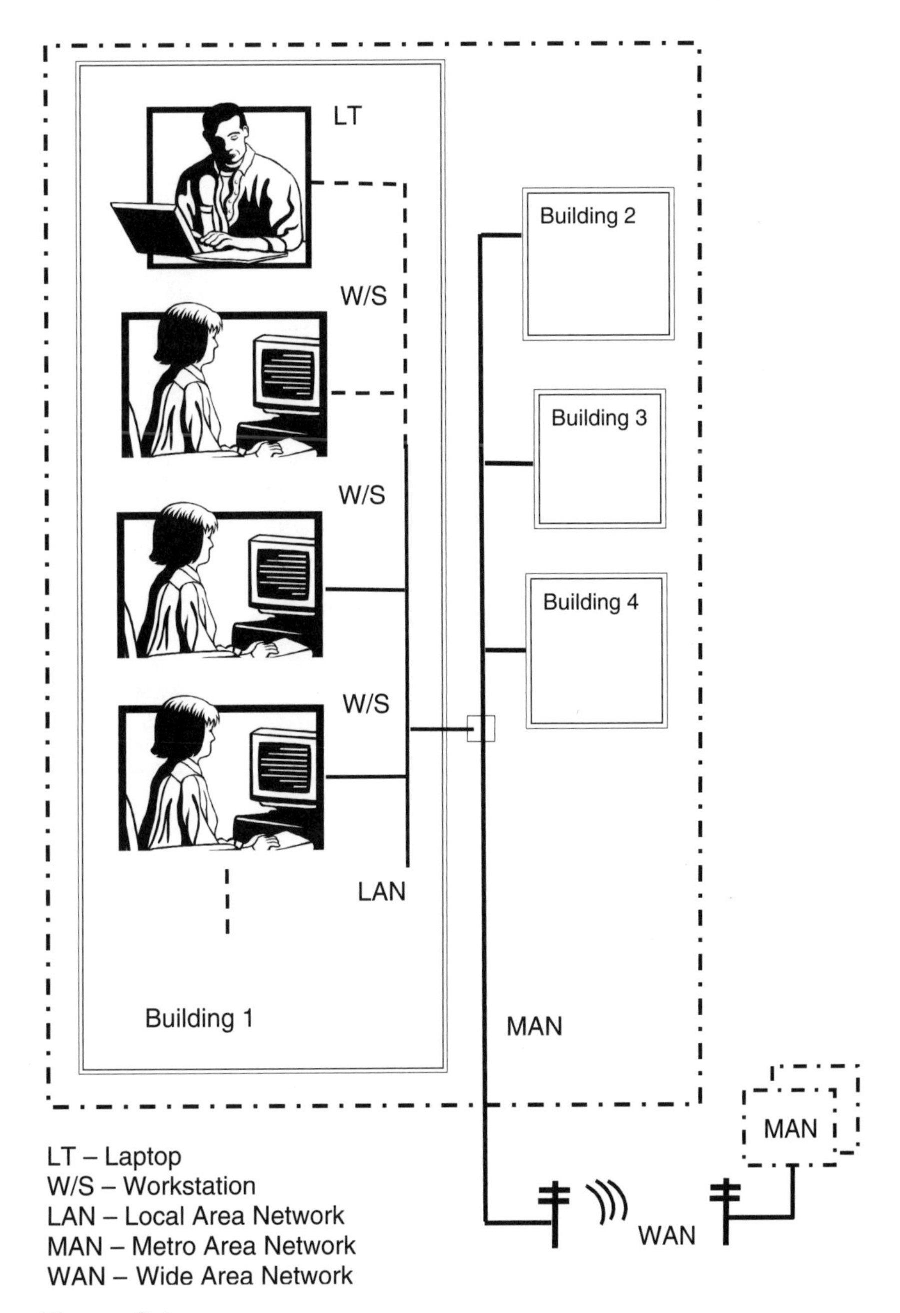

Figure 3.1
Computer Networks

Note that a *local area network (LAN)* links all the workstations in one building. A *metro area network (MAN)* can link several nearby buildings (as on a campus). The LAN or MAN can connect to a *wide area network (WAN)* to reach worldwide. The Internet is a public WAN. Some companies have private WANs.

Security Requirements

3. Third, what security requirements are required and are available? The crown jewels of the company will be on your computers. Your engineers need to work with unencrypted files. Industrial theft of *intellectual property* (*IP*) is quite common, unfortunately. It is not even a crime in some countries. (We mentioned the Cadence/Avant! lawsuit previously. IP includes design ideas about chips, tools, or anything.) Company attitudes towards security vary from clueless to paranoid. Some security examples in use are:

   Firewalls—these restrict who and what programs have access to and from the network.

   Dial-in *crypto tokens*—these have a user password and electronic token code that changes every few minutes. The codes must match the codes at the host computer.

   Frequent equipment location checks—location and content checkups of desk machines. Check-in and check-out of laptops.

Did You Know?

Laptops require extra security. Travelers can easily lose them or have them stolen. The very portability that makes them so handy also means that they can vanish in a second.

   Password control—this should be used on project files. The files should also be encrypted when they are not in actual design use. This is a real problem if a tool vendor needs your proprietary design files to resolve a tool bug. There needs to be *non-disclosure* terms in the tool contract (they promise not to reveal any sensitive user/customer information). The vendor also needs to have close control of who can access your design files at his shop. The vendor must ensure that the design files are not put on a networked machine with open access to non-authorized people.

   The security environment decisions may affect the computer environment choices, which is the next issue.

Did You Know?

Whatever procedures, encryption, firewalls, etc., are used to deter external access, the largest source (80%) of IP theft is your internal people!

Computer Systems

4. Fourth, what computer environment will be needed? All the previous questions affect this decision. On what computers do the current or expected EDA tools need to run (PCs, workstations, servers, or a group of servers)? Most large companies use farms of dedicated multiprocessor servers with huge amounts of memory and disk storage.

   On which operating system will the EDA tools run? Options include UNIX (most common), Linux (increasing), Microsoft NT (a few), or Windows (very few). There are different supported versions of each of these. The version differences may be minor, but may require compatible versions of EDA tools which share data.

   A few small EDA tools can run on Personal Computers (PCs). However, EDA users usually need high-resolution graphics, large screens, and fast processors. Most EDA tools require as much computer memory and disk space as you can get. They need long dedicated (uninterrupted) run times. As you know, there is great benefit from getting a product to market early. So, the designers should have access to the fastest possible computers.

Did You Know?

Some vendors offer accelerators (both hardware and software) that make the tools run faster.

Changing the computer environment is always a risky experience, with unforeseen delays. So, the managers need a flexible migration plan (and fallback plan!) to include upgrades, company growth (or reduction), and remote sites.

Nora: There certainly are many concerns to take into account. I was thinking it was just a computer decision. In marketing, I just use a laptop. What do your designers use?

Engineering / Non-engineering Goals

Hugo: You've touched a sensitive area. There is often a conflict between the company *Information Technology (IT)* and the engineering departments. The IT department seeks shared use of company central computers to lower costs, while the engineering department usually needs its own dedicated machines.

Conflicts arise over computer purchases (e.g., central vs. department servers, workstations vs. personal computers). Operating systems (NT vs. UNIX vs. Linux, etc.), and company-wide standards are also issues. The biggest argument may be over ownership / administration (i.e., company IT vs. local engineering).

Minimizing costs and support staff is the IT department's (in-house or contract) normal goal. Standardizing maintenance and centralizing all business applications are two ways to achieve this. High utilization of their computers (90% or more) is another.

IT business applications typically use small files (a few megabytes) and have short (seconds or minutes) transaction-oriented run times. User response time is not a major issue. User downtime and lockouts for maintenance are normal side effects. So are disk storage limitations. However, these are only annoyances for most of the organization.

In contrast, EDA tools operate with gigabyte files and long run times (hours and days!). Engineers make a small design change and re-run a whole suite of programs. Thus, long run times are the norm, not the exception. The engineering users need all the computing power and disk speed they can get. Moreover, a hangup or disruption is very costly to their project.

Most large workstations require administration and support (setting up and managing file storage, backup, network connections, etc.). These are usually network administration jobs, not IC designer tasks. So EDA users need IT support staff. They also need EDA support staff to manage the expensive tool licenses, upgrades, and usage.

EDA users need to **reduce** computer loading (utilization) to gain speed for their compute-hungry EDA tools. To meet design deadlines, engineering management must control computer priority and access to the tools and files.

Companies often address these conflicts with separate local computers, administration, and support for engineering. The rest of the company uses a shared IT server and administration. In companies with multiple projects or engineering departments, the same issues occur among the engineering groups. The engineering managers must share the high-powered engineering server farm. (I have seen these issues in several large companies.)

Nora: That solution must be difficult to sell.

Hugo: Yes, but it is essential if the organization wants to get the revenue benefit from the EDA tools.

Did You Know?

EDA workstations should always be fully equipped. The cost of adding memory, disk space, and high-speed network links is usually negligible compared to the costs of engineering time and TTM.

Nora: How do you go about buying EDA tools?

HOW TO BUY EDA TOOLS—FIVE KEY ISSUES.........

Hugo: First, you have to know the IC architecture that will be used—custom, standard cell, FPGA, etc. That narrows down the tool choices. Then there are five key issues to consider when buying tools. They are cost/performance, training/support, make or buy, compatibility, and transition. Let's look at them one at a time.

Cost/Performance

Hugo: How will the tool perform compared to similar tools? Tool demonstrations at the vendor's site, at our site, or at a conference are worthwhile. They can answer a lot of questions and raise some more. Of course, vendor demonstrations always show off the tool well (no surprise). They usually are done using small examples or to show just a few features.

Demonstrations also help to show up features from one vendor that are missing from the other. It's also good to ask each vendor why they are better than the competition.

Then you have to decide which features are most important to you. There are often several tools for a specific problem area, such as power estimation. Therefore, comparing the tools can take a lot of homework and engineering time.

Following up references is another way to gain insight on the tool performance. User groups are independent reference sources. For example, on the *Synopsys User Group (SNUG)* website, users describe tool problems and solutions. These include both Synopsys tools and others as well.

Unfortunately, some tool license agreements restrict dissemination of benchmark results. Vendors claim that results taken out of context do not provide a valid measurement. Therefore, users need to get comparative information informally.

What is the total cost for the tool? (Prices have varied from free university tools to over $750,000.) Tools may be bought outright, or licensed in several different ways. To compare different vendor tools you need to look at the total costs and the ROI to the user company.

What are the tool *licensing* options? These may be on a per engineering "seat" or per use basis or a floating license (any "seat" can use the tool). They also vary on a per month, quarter, year, or perpetual basis, and may be site-specific, company-wide, or worldwide. Table 3.1 shows some licensing options.

Table 3.1 Licensing Options

FACTORS	POTENTIAL OPTIONS
LOCATION	Department, site, multisite, division, city, country, worldwide
WHO	Named person, fixed seat (machine), floating seat, # of seats, employee, consultant, subsidiary, customer, etc.
DURATION	Free trial period, month, year, perpetual, subscription (quarterly)
PORTABILITY	Fixed seat, registered persons, dynamic person assignment, anywhere, anytime
CHARGE BY	Seat, minute, hour, monthly, yearly, usage, tool run, project, location, etc.

Hugo: Different vendors offer different variations—and new options emerge all the time. The table gives some examples of the variables possible. The costs may vary widely. For instance, a perpetual license may cost four to five times more than a subscription license.

A design manager needs to have as much flexibility as possible. Workstations fail or become obsolete, people come and go in projects and companies. Design managers need to be able to assign a temporary license to anyone they choose, as required. This may be anywhere, on any project, for whatever time they need. EDA licensing agreements can be a substantial burden for design managers.

The user should get a free trial period and automatic warnings **before** a license expires. The design manager needs easy license renewal, without either user or vendor bureaucractic delays. There have been cases where a license expired at a critical time without warning and took days to renew.

Some vendors charge for the right to use the tool, whether or not you are actually using it. It is obviously preferable to be charged only when the tool is actually being used.

The total cost involves the initial cost, the technical support cost, bug fixes, and upgrades. Yearly support cost often runs 15% or more of the initial cost. (All these things are usually negotiable...) In addition, there is your own in-house support cost to install upgrades and check bug fixes.

What is the cost/performance lifetime ROI? Will this tool create a better ratio of revenue improvement than a similar tool?

Nora: I see you need to take into account many cost factors.

Hugo: Yes, these are expensive tools, and there are a lot of them. If they make basic cost/performance sense, then we look at the support issues.

Nora: What kind of support issues?

Training and Support

Hugo: EDA tools are complex, with many options and parameters to set or enter. There is a huge amount of data to be entered in specific (often proprietary) data formats. *Graphical User Interfaces (GUIs)* help somewhat, but they differ from tool to tool.

Therefore, engineers need training on how to use the EDA tools. There is a substantial learning curve to become skilled in using each tool. Furthermore, each (frequent) upgrade of the tool may require a refresher course.

Did You Know?

EDA tools involve a constant learning process. Users often have to learn to use new tools or new features. Training time is a significant burden (losing a designer for a week or so) for any size company. Training can be quite expensive. Nevertheless, companies cannot afford to fall behind, and they need to train multiple people on each tool, to avoid dependency upon a single person.

Therefore, an important question is: what training is available? Are both initial and follow-on training included? Is emergency (24-hour hotline telephone and/or on-line) service provided? How good is the documentation? Program documentation is often minimal, particularly early in the life of an EDA tool. The vendor is focused on getting the tool up and working, so the documentation usually comes later.

How long will the vendor company be around to support the tool? Is there a backup provision to get the source code and documentation if the vendor support ends? EDA companies may fail or be acquired or merged. Their tools may be discontinued after such an event. Users then have to provide their own support or cope with a new tool and learning curve.

Will you get the tool source code (so you can modify it yourself if need be)? Perhaps you will get the tool object code (cannot modify it), or license the tool. (You normally don't want to touch source code if you are getting support and licenses from the vendor. This becomes more of an issue with an unsupported or poorly supported tool.)

Is there a schedule for updates and releases from the vendor? How and when will change control be implemented in your shop? Who will do it —an engineer or an administrator? When will it be updated—at regular intervals or at specific points in the design?

Nora: It sounds like a long-term relationship is necessary with the tool vendors. Do you buy all your EDA tools from vendors?

Make or Buy

Hugo: Yes, we use only vendor tools. It's usually the easiest and fastest way to go. Most companies use vendor tools. However, some companies develop their own tools when they are not available from a vendor.

"Make or buy" is another important EDA user decision. If done in-house, the company must commit to long-term support. Just as with vendor tools, internal customers want more support and features for the in-house tools.

We also buy tools to help with essential tasks that no one likes to do. These include design documentation, user manuals, and change notices.

Nora: Is there any problem with compatibility between in-house and vendor tools?

Compatibility

Hugo: Yes, there is a big issue of compatibility between vendors tools and between vendor and in-house tools. How compatible is the tool with our existing suite of tools? (Who will make the interface translators if there is a compatibility issue?)

Will there be *backward compatibility*? (Will upgrades of the tool be able to run old versions of the design files?) This can be key if revisions to a product design are made (usually the case). Some managers resist installing tool upgrades to avoid this kind of problem.

On the other hand, vendors will not fix bugs on two-year-old versions (no surprise). And managers cannot easily move people from one project to another if each group uses a different tool version. So companies must upgrade versions frequently, and all users together! This is often a major complication.

Nora: I thought that was a problem only when they upgrade the operating system. You mentioned earlier that the tools have to match the operating system version.

Hugo: These are issues with both tools and operating systems. In addition, improvements in the tool may often be incompatible with prior design work. Developers tend to expect you to use the new tools only on new designs. You must keep old tool versions available until you are certain they are no longer needed.

Nora: The transition to using a new tool or upgrade sounds tricky.

Transition

Hugo: Well, you need to be sure that an engineering group/project is willing to be the "first user" for a new version of a tool or operating system.

I mentioned earlier the update schedules from the vendors. What is their timetable for bug fixes and/or upgrades for a new tool? (A new

EDA product will need more productization to get it usable for a real design.) When will it be ready for your first user and will the vendor give extra support?

Is there a Trial Period and Escape clause (i.e., if you try it and don't like it, can you get a refund)? The vendor is unlikely to negotiate a low price with such a clause. But you need an agreed-upon *acceptance test* as a criteria for acceptance or rejection.

The internal staff also serves as user gurus for the design engineers for all the EDA. If new tools are bought, will you need more in-house staff to support them? More staff can be a budget issue unless the EDA tools are shown to directly support revenue-generating projects.

Hugo: So those are the five essential issues in buying EDA tools. I hope you are not overwhelmed. See the little picture I have over here that summarizes the negotiation process. Everyone has his or her own focus area, but we have to cover all the concerns. (See Figure 3.2.)

Figure 3.2
Buying EDA Tools

Nora: That's a good summary. There sure are many factors to negotiate. Do the marketing people from the tool vendors assist you in that?

Hugo: Certainly, up to a certain point. Even with reference users, we always have to try the tools ourselves. We have to ensure they are really compatible with our existing tools and design needs.

Nora: Yes, someone else mentioned the user's tool integration issue. What is the story on that?

STANDARDS EFFORTS—WHO, WHAT, AND WHY ...

Design Flow Integration

Hugo: The problem is this. The IC design engineers seek to assemble the best-in-class tools for their design system. This means picking one tool from vendor A, another tool from vendor B, and so on.

Most EDA tools read in one or more *design files* (file of design data). The tool or the designer may generate additional data and then write out an enhanced version of the design file(s).

Each tool arranges the file data in a specific way (*format*) for fast operation. The different tool functions are generally sequential, with the data passing from one tool to the next. (However, recent tool suites are trying to do more things concurrently.)

Most EDA tools are developed independently of each other. Design groups use various scripting or programming languages (such as Scheme, SKIL, or Perl) to stitch the tools together into a *design flow.* A design flow is the sequence of specific design tools used.

These scripts translate the data from one tool format to another, call up the correct files, and initiate the EDA tool operation. The scripts provide the glue that simplifies the tool design flow for the users. They also reduce the amount of tedious, error-prone manual work in running and using EDA tools.

Tool integration scripts would be easier to write if tools communicated in a standard way. However, most tools were developed independently with little concern for a standard *interface* or file format. (There is no standard scripting language, either, by the way.)

An analogy would be different paper filing systems. Suppose I keep a file of newspaper articles organized in three folders by date, author, and newspaper. Suppose you keep a file of articles all in one folder, arranged by date, subject, author, and article length.

I wouldn't know how to find something in your files, and you wouldn't know how to find something in mine. However, some of my articles would be of interest to you and vice versa.

With data files, the *data representation* (numbering, units, location, coding) may also be dissimilar. Here, I will sketch you a picture of what I mean. (See Figure 3.3.)

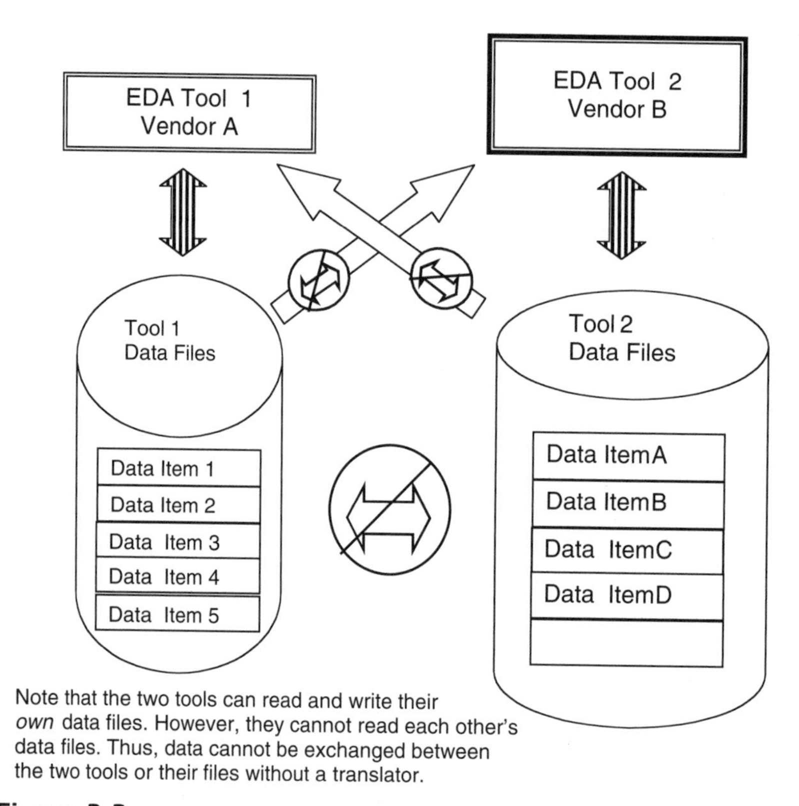

Figure 3.3
Lack of Tool Interface Standards

Tool 1 data files may have data items such as:

Name: Gate31

Netdelay: 0.3

Location: G78

Block #: 300

Chip #: ACD&

Tool 2 data files may have similar and dissimilar data items such as:

G78

Gates—31, 33, 36

Dly—300 psec

Revision—88

The number and order of items and the specific data format in each file are different, and each file may have data which the other does not.

So, as you can see, even though the data items contain common information, EDA Tool 1 cannot read the data file of Tool 2. A utility program or script is required to translate the data.

Nora: Are there no standards?

EDA Tool Interface Standards

Hugo: Yes and no. There are some. Standards help ease the task of assembling a workable EDA *design flow.*

Tool users or tool vendors, or both, usually create standards. There are always issues of vested interests, politics, resources, and competing standards groups. Other issues include testing conformance to the standard, vendor and user adoption, and so forth. Standards take years to create and may be obsolete by the time they are done!

Some standards are created after the industry fights over competing approaches. Then they settle on a standard approach, although it might not always be the best technical solution.

Other interfaces become *de facto* standards simply by broad acceptance and use. An example of a *de facto* standard is the *GDSII (Graphic Design Station II)* format. It is used for the graphical layout data of IC masks. (However, there is an effort to replace it with an improved and more compact standard format.)

An example of a planned EDA file standards group is the *Electronic Design Interchange Format (EDIF).* This is a human-readable file format developed to enable cooperating semiconductor manufacturers to exchange data. Although not intended to be an EDA tool interface, it has been widely used for that.

Vendors who offer a suite of interoperable tools also provide interfaces (often proprietary) between their tools. They may support an interface to **read** data in from other vendors' tools. This would bring the design data (and the customer) into **their** tool suite.

However, it is **not** usually in their interest to **write** data from their tools to a competitor's tool. They might lose the customer. Therefore, although many vendors claim EDIF compatibility, it has been mostly **read-in** only. (The company Engineering DataXpress has long provided translators and other services based on the EDIF standards.)

Nora: So there are "semi-standards"?

Hugo: I guess you could call them that. It's a little like railroad tracks. Historically, different railroads used different types of tracks. Each railroad "standardized" on their type of track. However, a train could not move from one company's track to another! Passengers had to switch trains.

In the 1980s, the industry really tried to create a universal interface.

Frameworks

The ASIC Council is a small (seven or eight members) consortium of the largest ASIC semiconductor manufacturers. In the 1980s, the council sponsored the Silicon Integration Initiative, Inc. (SI2). It worked on a universal *framework* into which users could easily plug in different vendor tools.

However, the result was multiple vendor proprietary frameworks, instead of a single framework standard.

Nora: So standards have not always succeeded. Is there at least a common database for all the EDA data views?

Design Database Standards

Hugo: A good question. The answer is—not yet, but that may be coming. Usually each tool's files have an internal arrangement of data items (*data structure*) optimized for that tool. The data structure affects how fast the tool works. It also affects how easily different tools can (or cannot!) exchange data.

Some vendors have a proprietary database to which all **their** tools can talk (*interface*). As more multivendor tools need to work together, a **common** design database becomes more important.

An open, universal database would be a great improvement for users. There are significant technical and business implications, making this a difficult standardization goal. Several existing databases or access methods are being considered in a couple of standards committees.

Nora: Are there several standards groups?

Standards Groups

Hugo: Yes, many of them. With new ones forming all the time, some compete with each other. As one industry pundit said, "The best thing about standards is that there are so many of them!" I'll mention just a few groups:

- The Institute of Electronic and Electrical Engineers (IEEE). Primarily for hardware designers, it has developed hundreds of standards.
- The Joint Electron Device Engineering Council (JEDEC), a division of the Electronics Industry Association (EIA). It has electronic product standards of all kinds (48 different committees).
- The Special Interest Group on Design Automation (SIGDA) of the Association for Computing Machinery (ACM). It is for EDA professionals. (In spite of the name, this is a software group, not hardware.)
- The Virtual Socket Interface Alliance (VSIA). For product, IC, and EDA vendors, it creates standards for system-on-chip *Virtual Components* (blocks of intellectual property).
- Accellera—an EDA and designers' group. It is trying to standardize a formal verification language and a system design language. Accellera manages two prior standards groups, Open Verilog International and VHDL International.
- Open-Access Coalition (OAC). Led by Silicon Integration Initiative (SI2), this user-backed group has established an industry-wide data model and application programming interface (API) which potentially can access any database.

Nora: Do your people work on any standards committees?

Hugo: Sometimes, but it takes a lot of (unpaid) time. We get involved when we need a standard to have a salable product. Sometimes we sit in just to ensure that a competitor doesn't dominate the standard to fit its product!

Nora: Can you tell me more about your EDA staff?

PERSONNEL—THE KEY TO EDA SUPPORT

Hugo: That's a good area to discuss. An essential element in the success of a design group is the EDA support staff. It administers the EDA design system with its tool licenses, libraries, computer, upgrades, and backup support.

In any company, there is a significant learning curve for new people. There is a need for job satisfaction and incentive for experienced people. The company needs to cross-train people. If one person leaves, that should not remove the only knowledge about a particular area or tool.

Most EDA staff members want to work with the newest tools—to develop, test, or use them. However, a large part of the workload is the support and maintenance of the existing tools. EDA managers try to balance this research-and-development (R&D) vs. maintenance dilemma in several ways.

One approach has junior people learning the tools and problems by handling most of the support. They may move on to R&D after they learn about the existing tools. However, experienced staff also needs to be available to train them in the support work.

A second approach rotates the staff through both roles—so everyone does some R&D and some support for a time. This must be done carefully to avoid losing continuity in either area.

Another approach has every R&D person also supporting one or more tools. That way they have a role in, and appreciation for, both areas. They also learn how to make their development work easier to maintain.

Nora: That's quite a balancing act. Where do you get your staff?

UNIVERSITY CONNECTIONS...............................

Hugo: From other companies, of course, and from colleges and universities. Many universities have industrial liaison programs to present their research projects to industry. Often EDA companies fund the research and hire graduate students as interns.

We hire quite a few new college graduates—usually former interns. We maintain contacts with several leading universities which are very active in EDA research. Most of these have strong industry relationships as well.

Did You Know?

Many EDA companies (and large ASIC and FPGA vendors) provide low-cost licenses and training to universities. Vendors want the students to become familiar with their product. The students are more likely to use it at work later if the product is familiar. And their companies like the shorter learning time.

These universities are involved in EDA research, and their professors consult to the industry. Many EDA companies have technical advisory boards that meet regularly. They review the research products and technical direction of the company. Their input is critical to the company for future direction and to avoid noncompetitive products.

Internship benefits the student in learning how the real world works. It also provides low-cost expertise for the company. Many student interns return to work full-time and can contribute immediately with little learning delays. I can give you a list of some universities with which we work.

The list includes:

Carnegie Mellon University

Massachusetts Institute of Technology

Princeton University

Stanford University

University of California at Berkeley

University of California at Los Angeles

University of California at San Diego

University of Illinois at Urbana

University of Texas at Austin

And many others.

(Appendix E has websites and other information on these universities.)

Nora: Thanks, Hugo. Now I have a better appreciation of our EDA customers' problems.

SUMMARY

There are four important new EDA user questions:

What is the design organization?

What kind of network capability is needed?

What are the security requirements of the computer network?

What computer systems are needed?

There are several considerations in buying EDA tools.

One is cost versus performance. Another consideration is the choice of licensing options. Training and support are critical factors, as is compatibility with the user's existing tools. Finally, there are concerns about ensuring a smooth transition in using the new tool.

Integrating tools into a design flow is not a simple task. Many EDA tools and data files are not compatible with each other. There are many EDA standards groups trying to improve this situation.

EDA staff comes from universities, electronic product and IC manufacturers, and design houses. Universities have close financial and technical relationships with industry.

QUICK QUIZ

1. Which one is **not** a key **initial** decision for a new EDA user company?
   a. Design organization (who and where)?
   b. Network capability needed?
   c. Time-to-Market?
   d. Security requirements?
   e. Computer system needs?

2. Which is not an important factor in buying an EDA tool?
   a. Runs under Windows
   b. Offered by major vendor
   c. Cost/performance
   d. Have staff already trained on the tool

3. Most vendor tools are compatible with other vendor tools.
   a. True
   b. False
4. EDA tools training is usually not required.
   a. True
   b. False
5. EDA tools usually run quickly and use little memory.
   a. True
   b. False
6. There are many types of EDA tool licensing options.
   a. True
   b. False
7. Which of the following is an important standards group?
   a. IEEE
   b. Accellera
   c. VSIA
   d. All of the above
8. Design Flow refers to
   a. Fast or slow design speed
   b. Style of design
   c. Sequence of EDA tools used
   d. Flowchart design
9. Much university EDA research is supported by:
   a. IC manufacturers
   b. EDA companies
   c. Government agencies
   d. All of the above

Answers: 1-c; 2-a; 3-b; 4-b; 5-b; 6-a; 7-d; 8-c; 9-d.

4 Overview of EDA Tools and Design Concepts

In this chapter...

- Introduction
- Major Classes of EDA Tools
- Essential EDA Concepts
- Design—The Art of Trial and Error
- Architecture, Methodology, and Design Flow
- Summary
- Quick Quiz

INTRODUCTION

In the next few chapters, we will discuss specific kinds of EDA tools. In this chapter, we provide an overview of the EDA tool world.

There are several EDA-related concepts which are helpful to understand the EDA tools and the design process. These include:

- Design Views
- Hierarchy
- Design
- Chip Architecture
- Methodology
- Design Flow
- Tool Suites

Nora of Sandbox, Inc., is meeting again with Andrea, an ASIC engineer who works for Fabless Design, Inc. Nora has asked to learn more about the use of EDA tools. Let us listen in on their discussion.

Nora: Hello again, Andrea. Did you like the party the other evening?

Andrea: Yes, very much. Sandbox has both outstanding tools and great parties.

Nora: We seem to be continually developing new tools. I am a little unclear why that is.

Tool Improvements

Andrea: That's because each generation of semiconductor process improvements shrinks the size of the transistors and wires. The number of transistors and wires on the chip increases dramatically.

The wires and transistors are smaller and closer together. This increases the electrical coupling between transistors and wires, which can cause more failures. The chips run at higher frequencies, which leads to interference on the wires. Power increases with faster chips and that can cause thermal problems as well.

All of these issues cause new kinds of design obstacles. New EDA tools are created to address these design challenges. Tools also need more capability just to handle the increased numbers of transistors and wires.

That is why developers are constantly updating and creating new EDA tools.

MAJOR CLASSES OF EDA TOOLS

Nora: There are so many EDA tools. Is there a grouping or order to them?

Andrea: Yes. There are three major groups or classes of EDA tools. They follow a general design sequence. Let me show you this overall sequence. (See Figure 4.1.)

Note the three main groups: *electronic system level (ESL)* design, IC *front-end (FE),* and IC *back-end (BE)* design. Each consists of design tasks, followed by verification or checking steps. If the checking steps find errors, the design is revised. The designers repeat (iterate) this design-verify loop as needed. The iteration may even go back to the front-end or system design requirements if back-end errors are found.

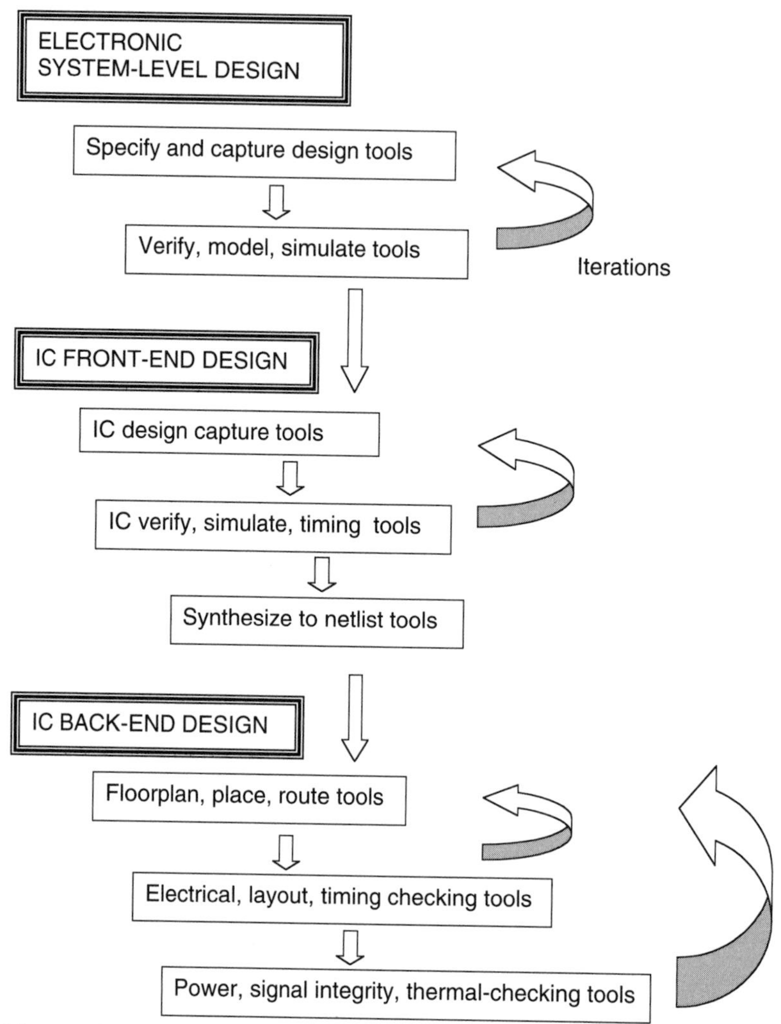

Figure 4.1
Overall Sequence of EDA Tool Use

Electronic System-Level Design Tools

Andrea: *ESL design* uses one class of tools. Just as the house architect describes the customers' requirements, so does the electronic product engineer.

The architect builds a model (on paper or on the computer) to see how the design will look. The system engineer also makes a high-level computer *model* to see if the design will work.

The architect may try several models, exploring different materials or house layouts. The system engineer does *design exploration*, trying different design approaches (such as implementing functions in hardware or software).

Nora: So there are EDA tools to help with defining requirements, modeling the system, and exploring different design approaches?

Andrea: Yes. Most of an electronic system can now fit on a chip. Therefore, IC designers and system designers now need to know about both system design and IC design!

Traditionally, IC design falls into two major divisions. These are the front-end design and the back-end design.

Nora: So we have system, front-end and back-end tools. At the party, I think I talked with engineers who used each kind. What are the front-end tools?

Front-end Design Tools

Andrea: *Front-end* IC design tools include *design capture*, *verification,* and *synthesis* tools. A house architect records design ideas on paper or on a computer. IC designers similarly capture their design ideas on a computer, using words or graphical symbols (design capture).

The next task is to *verify* or test that the design ideas look right or work correctly. For this, architects and engineers used to build detailed miniature models (*prototypes*) of their designs.

However, now computer programs can *simulate* almost anything. Simulation is faster and less expensive than prototypes. Architects can present a simulated walk-through of the house. (You can even see this in real estate ads on the Internet!) IC designers can simulate every operation of the IC, from input to output.

IC *design verification* includes simulation tools to verify that the design behaves and performs as intended. Additional *analysis* and *checking* tools ensure that the design meets all the *design rules*. (Design rules are like the electrical and plumbing codes which cities require for safe buildings.)

The third major front-end task is *synthesis.* An architect transforms the house design into a list of specific parts, materials, and building instructions. IC synthesis transforms the design description into a specific set of components and wire connections. A *synthesis* program selects the transistors, gates, or cells from a library.

The synthesis output is a *netlist* file of transistors, gates, or cells and connecting wires *(nets).* This is the dividing line between the front-end and back-end design steps.

Nora: Okay, front-end tools help with design capture, verification, and synthesis. What are back-end tools?

Back-end Design Tools

Andrea: The *back-end* design tools are collectively known as *physical design* tools. They aid with the general layout of the chip *(floorplanning).* They also help the detailed *placement* (locations) of the transistors, cells, or IP blocks on the chip.

After placement, *routing* tools *(routers)* help plan the physical routing of the interconnect paths.

Timing-check tools calculate delays with the new information *extracted* from the actual physical layout.

Other tools check the many *electrical* and *physical design rules.* Most physical design rules relate to manufacturing tolerances. ("The space between wires may not be less than 2 microns" is an example of a physical design rule.)

The output of the back-end design flow is a *tapeout* file. It contains the graphical patterns (*polygons*) for each transistor and wire on the IC.

Nora: So we have floorplanning, placement, routing, extraction, and rule-checking tools in the back-end physical design?

Andrea: Yes. Let me show you a table of typical tool types and who uses them, in each area. (See Table 4.1.)

Table 4.1 Major EDA Tools

ELECTRONIC SYSTEM-LEVEL (ESL) DESIGN	
Users	Tools
Architect	ESL Design Entry
System Engineer	ESL Verification
IC System Engineer	ESL Modeling
	ESL Timing Analysis
FRONT-END (FE) DESIGN	
REGISTER LEVEL	
Users	Tools
	RTL Entry
System Engineer	Test Bench
Logic Designer	RTL Simulation
ASIC Designer	Formal Verification
Test Engineer	Design for Test
	Timing Design
	Thermal Design
	Power Design
	Signal Integrity Design
	Synthesis to Gates
GATE LEVEL	
Users	Tools
Logic Designer	Schematic Capture
ASIC Designer	Gate Level Simulation
BACK-END (BE) DESIGN	
Users	Tools
	Floorplanning
	Layout—Place & Route
Layout Designer	Electrical Rules Check
Logic Designer	Physical Rules Check
ASIC Designer	Extractors & Delay Calculators
Test Engineer	Timing Analysis
	Power Analysis
	Thermal Analysis
	Other Analyses

Andrea: The table summarizes what we have been talking about. You see the three main areas, the primary tool users, and the major tool types. The two divisions (FE and BE) overlap now as more physical design and estimation is done up-front. (We haven't talked about all these tools yet.)

Other tools not in the three main areas address documentation, change, and version control. Chip packaging also has its own suite of design tools. So do analog, radio frequency (RF) and mixed-signal design tools. (See Appendix C for more information on these.)

Note that front-end design can be done at two levels. Tools work either at the detailed *gate level*, or at the higher *register transfer level (RTL)*.

ESSENTIAL EDA CONCEPTS

Nora: With all these tools, there must be a lot of data around. Would you please explain more about the EDA design data?

Design Views

Andrea: There is a lot of data. That's because an IC design is described from many different viewpoints. Just as in the case of a house, different people need different views.

The prospective buyers may want to see the arrangement of rooms (the floor plan). They may also like to see an exterior view (artist's drawing), including landscaping, fences, and so on. A glossy brochure will typically contain these views plus a list of features.

A more detailed block diagram includes the location of all the closets, appliances, bathroom, and kitchen fixtures.

The plumbing contractor needs a detailed plan showing the size and location of each pipe. The electrical contractor needs a wiring diagram showing each cable and outlet. The customer or the interior decorator doesn't want to see those views.

Nora: Yes, I understand the need for different views.

Andrea: Similarly, an IC design has different *design views*. The system designer works with a description of all the functions and features needed.

A logic designer needs to see all the gates and logic displayed as pictures *(graphic symbols)* or described in text.

The layout designer needs to see a graphical floorplan of every block and transistor in the design.

One difference between a house and an integrated circuit is that the house is passive. It just sits there (except in earthquake zones where it might slide down a hillside!). The house plans are *physical views*.

The analogy is better if we include activities of the people living there. They move from room to room, watch TV, read, cook, and sleep at specific clock times. They perform specific functions at different speeds. We could describe these activities as *behavioral, performance, and timing views.*

An IC has electrical signals and data moving around. It exhibits behavior, operates by clock times, and runs activities at different speeds. Therefore, it also has physical, behavioral, and performance (timing) views.

Nora: I imagine there are even more views.

Design Data

Andrea: Yes, there are, and more kinds of data as well. In houses, there are *design rules* and guidelines. Building codes require minimum pipe and cable sizes, safety features, and materials to be used. *Design guidelines* suggest things that should be done, or should not be done (*constraints*). "Locate the dining room close to the kitchen" is a guideline example.

ICs also have a host of design rules and guidelines. All this different data also resides in the various views and is stored in files. Just as people organize their paper files differently, so do programmers. (With luck, they remember where they filed things better than I do!)

For instance, I may have a telephone list of friends, ordered by first names. Your telephone list may arrange them by last names. I may just include the telephone number, while you include the address. My telephone number format may be (888) 732-4567, while yours is 888 732 4567.

Therefore, the design *data representation* (data content, format, and arrangement) may be different for each view. A design consists of the whole set of views. An individual EDA tool usually works with one view and one design data representation.

Suppose the timing data from one tool's file would be useful to another tool. That data might have to be translated from one representation to another. The incompatibility of EDA tools is largely due to data representation differences!

Nora: Yes, Hugo mentioned the tool incompatibility problem. Is some data more important than others? Is there a hierarchy of data?

Design Hierarchy

Andrea: There is a hierarchy, but not because of importance. *Hierarchy* is a useful concept for houses, system, or chip design. It refers to describing a house, project, or chip as a collection of different pieces. Each piece is then described in terms of smaller pieces or blocks.

Most large projects are broken down into smaller chunks to make them easier to understand. People do that all the time.

For example, we can describe (or draw) the house as a two-story brick dwelling. We can describe it at a lower level as a collection of rooms (e.g., living room, bedroom, and kitchen). We can next describe a room (e.g., 10x14 feet square, three windows, and a closet).

Note this is a hierarchy of the physical view of the house. We can have a hierarchy in each kind of view. Like the house hierarchy, we have a chip design hierarchy of blocks, cells, gates, and transistors. Look, here is an example sketch of a typical design hierarchy. (See Figure 4.2.)

Note that the IC contains blocks contain cells, which contain transistors. Hierarchy reduces the visible complexity for the viewer. Each higher level of description hides the details of the lower levels.

All the design details are exposed when the data is *flat* (no hierarchy). With millions of transistors, that's a lot of exposure. A related term is *level of abstraction.* The higher the level of abstraction, the less detail is shown.

Nora: Can all EDA tools work with hierarchical information?

Andrea: No, some EDA tools cannot handle hierarchy and work only with flat design data. Flat design files are huge, since they contain all the detailed data. So those tools run more slowly. Hierarchy can enable smaller design files, so the tools can run faster. Another point is that the hierarchical levels do not necessarily match across the different views. Thus, physical view design levels may not correspond to the logical view design levels.

Nora: All right, I see that design data comes in many views and can be hierarchical. Can you explain just how you design something?

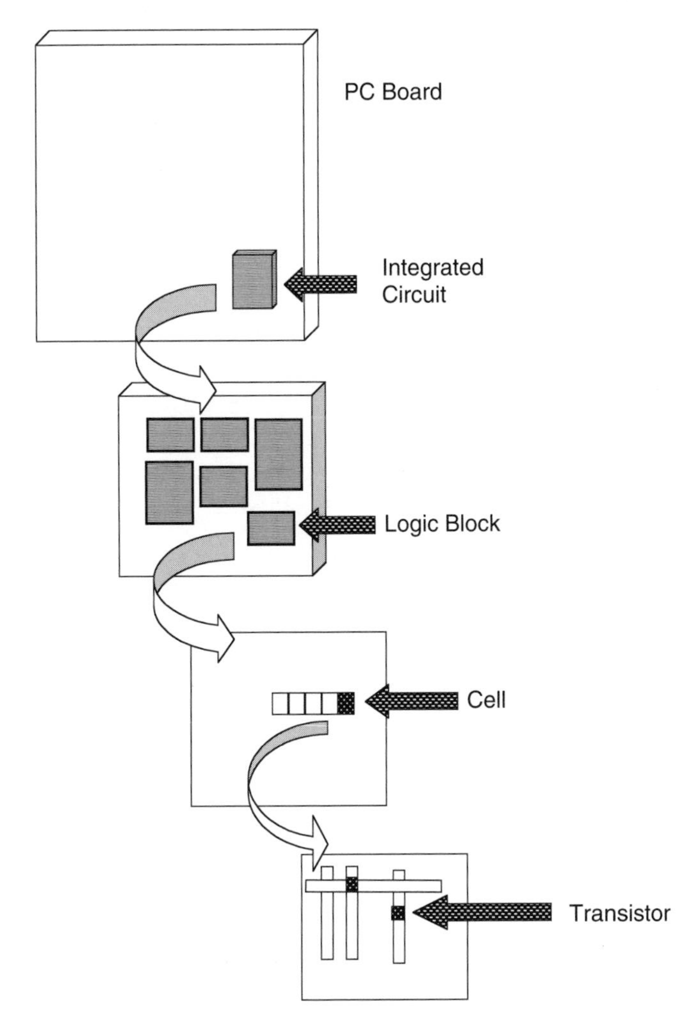

Figure 4.2
Hierarchy Example

DESIGN—THE ART OF TRIAL AND ERROR

Andrea: That's a good question.

Suppose we want to build a doghouse. We could build it from existing things like pieces of lumber, roof tiles, screws, and paint. We plan what it will look like and how big it should be. We decide what materials to use, how much to buy, and what it will cost. The plan shows how all the pieces fit together. What are the steps needed to build it, and in what order do they have to be done? The step-by-step procedure of how to build it is part of the plan.

This planning process is called designing and the resultant plan is called the design.

An architect starts with client requirements and makes a plan or trial house design. The plan may use existing parts or materials used as in prior designs or in a new way. If necessary, the architect creates a new part (a new shape of window, for instance). There are many types of styles and materials from which to compare and choose. The architect must make *trade-offs* among cost, function, and appearance.

Similarly, electronic design includes defining the product requirements, creating trial designs and exploring design trade-offs. The system engineer must trade off issues such as cost versus performance, or hardware versus software.

The designer breaks (*partitions*) the trial product design into smaller chunks. These may be new or re-used design blocks, either in hardware or software. Blocks may be separate functions or physical blocks (e.g., a voice recognition block or a microprocessor).

Designs are usually incremental advances to something that already exists. Modifying an existing design to make it smaller, run faster, or use a new interface are examples.

Design engineers usually don't get it right the first time. They make a trial design, then *verify* (test or measure) it to see if it works. If not, they fix the error, and try again.

The more complex (or new) the design is, the more trials are needed to get it right. Each trial and error sequence is a design iteration. The design team must create both the design and the tests, so errors are likely in both.

Did You Know?

Verification **is the process of confirming that the design does what the designer thought it should do, i.e., the design works as the** ***designer intended. Validation,*** **on the other hand, is the process of confirming that the design does what the** ***customer intended.*** **These two terms are often confused, or used interchangeably, or used together as a pair, i.e., verification and validation.**

Nora: Okay, so you redesign repeatedly until you get it right?

Andrea: Yes, and we learn more about the problem on each pass. The description of the customer problem or goal may be ambiguous, fuzzy, or incomplete. Getting the customer's requirements (the *specification)* correct and complete is one of the hardest design problems.

A customer often has only a vague idea of what he or she wants or needs, as strange as that seems. "Right" means complete, unambiguous, accurate, and possible to do. That's true when designing a house, an EDA tool, an IC, or an airplane. It applies to every level of design.

There are three main causes of design iterations: specification errors, design errors, and test errors. Better EDA tools help us get it right faster.

Nora: Okay, but how do you start a design?

Design Styles

Andrea: The designer may start from the highest level of *abstraction* (least amount of detail). Then the team repeatedly refines the design to lower levels of detail. This approach is called *top-down* design. Sometimes an engineer will keep the design very abstract and explore different architectures (e.g., gate array, standard cell, or software).

In a second approach, (*bottom-up* design), the design team starts with well-known low-level blocks and builds upward. Sometimes the team is constrained to use only existing parts for reasons such as cost or schedule.

A third style (*middle-out* design) starts with some known portion in the middle. The designers progressively describe the design up (in more abstract terms). At the same time, the individual blocks are refined down (in more detailed terms).

At other times, the design engineer will start a design based on known physical blocks or constraints (*block-based* design).

None of these approaches is fixed. There are variations of these styles, and some designs use a mixture of styles.

Nora: So how do you know which style to use?

Andrea: Experience, mostly, the constraints of the problem, and what the EDA tools will allow us to do.

Nora: But how do you decide who on the team does what?

Design Partitioning

Andrea: That's part of the *design partitioning* task. We break up any design job into partitions (blocks or sections). We may partition it to allow parallel work on different sections. Alternatively, we may partition it by function into software and hardware portions. Or we may partition by which parts can be re-used and which parts need to be designed.

In IC design, we also partition the design work by people according to their skills. In the IC front-end, we may partition by chip architects, logic designers, analog designers, and programmers. The back-end design phases include layout designers, verification engineers, and test engineers.

ARCHITECTURE, METHODOLOGY, AND DESIGN FLOW

Nora: What do designers mean when they speak of chip "architectures"?

IC Architectures

Andrea: That is more jargon. There are many ways to implement an IC design. The designer chooses a specific physical IC *architecture* to implement the design (e.g., GA, SC, FPGA, or custom). Do you know about those?

Nora: A little—someone at Sandbox said they were different ways to make an IC. He emphasized the time-to-market issue.

Andrea: That's right. Here, let me sketch you a picture of some IC architectures. (See Figure 4.3.)

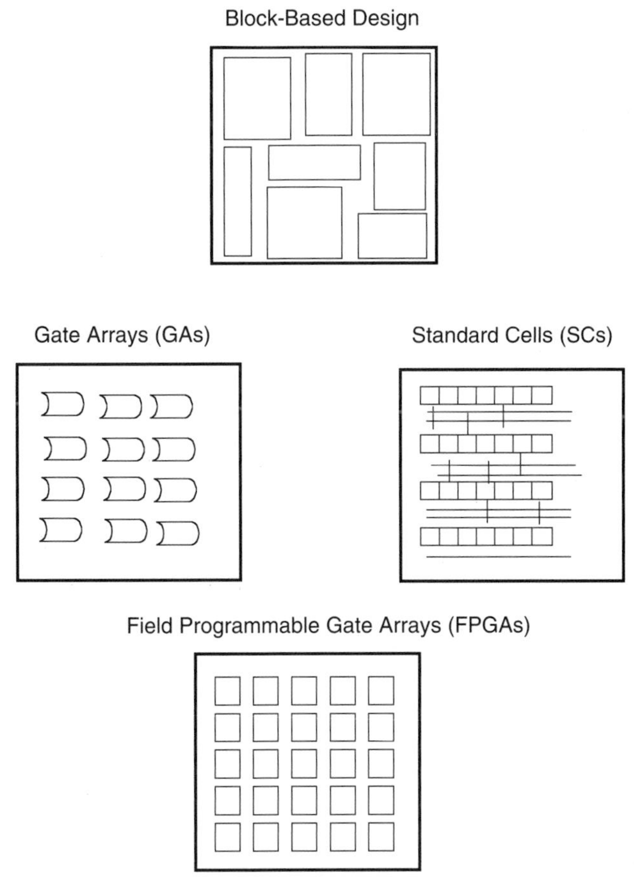

Figure 4.3
IC Architectures

Note the four different physical IC architectures. The block-based IC is usually a custom chip. The individual blocks may be custom logic, or memory blocks or microprocessors. The cells in the standard cell and FPGA chips may be many different kinds of gates. And FPGAs come in various flavors and styles as well.

The gate array is a fixed array, usually of the same complex gate. The gate array and FPGA chips come in fixed sizes determined by the manufacturer. However, the standard cell and block-based chip can vary as large or small as the application needs.

Also note that combinations of these architectures are common. For example, there might be some fixed microprocessor or memory blocks embedded in the FPGA. A standard cell implementation of some function might be just one block in a block-based chip.

Nora: As I recall, the gate array, standard cell, and FPGA were all ways to simplify the design and speed the time-to-market.

Andrea: That's right. We developed architectures and design aids for the most repetitive and tedious aspects in design.

Nora: What do the designers mean when they talk about "methodology" and "design flow"?

Design Methodology and Design Flow

Andrea: Let me show you a comparison of design methodology and design flow. (See Figure 4.4.)

Design Methodology (sequence of steps)

1. Design capture
2. Verification
3. Synthesis
4. Layout

Design Flow (sequence of tool operations)

1. Open design capture tool 3.1
2. Enter setup, configuration, and parameter information
3. Enter and edit design description (HDL)
4. Save design file "myfile-A"
5. Enter test vectors and save in "testfile-T"
6. Open simulator tool 2.2
7. Enter setup, configuration, and parameters
8. Run simulation with "myfile-A" and "testfile-T"
9. Display report(s)

Figure 4.4
Design Methodology and Design Flow

Note that *Design Methodology* is the sequence of design steps needed to implement a particular architecture. Design steps include entering a design into the computer, testing it, doing physical layout, and so forth.

If the physical architecture of the IC is different, the design steps may be different. For example, designing an FPGA requires different design steps from designing a standard cell IC.

A *Design Flow* is the specific set or suite of *EDA tools* used to implement the design methodology steps. The design flow is often a script calling out the tool and file details for each operation. Some of the methodology steps might involve more than one tool and multiple files. What I sketched here is a very simplistic overview. Real methodologies are much more complex and detailed.

Nora: Okay, I think I get the general idea. Hugo explained the business differences, but these sketches help. Can you explain more about tool suites?

Tool Suites

Andrea: Okay. At particular points in the design sequence, we developed individual tools to aid manual design. Different groups created the tools independently of each other and at different times. Therefore, most *point tools* have different user interfaces, data entry methods, and data storage styles.

Some point tools were modified to work together and were combined into clusters called tool *suites*.

Nora: Thanks, Andrea.

SUMMARY

There are three major classes of IC EDA design and tools: electronic system-level, front-end and back-end. ESL design includes requirements definition, modeling, and design exploration. FE tools include design capture, verification, and synthesis. BE tools include place and rout, as well as electrical, logical, and physical design rule checks.

IC designs are described from several different viewpoints, including behavior or functional, performance, and physical views. Each view consists of different kinds of information or data. Each view may contain several levels of hierarchy. Each level of hierarchy hides details of the levels below it.

Chip designing is the process of planning an IC and refining the plan. The resultant design is verified and redesigned until it meets the design goals.

Design methodology is the sequence of design steps followed for a particular IC type. A design flow is the series of actual tool operations used to implement a methodology.

QUICK QUIZ

1. System-level design includes:
   a. Design modeling
   b. Place and route
   c. Synthesis
   d. Design rule checks

2. FE design includes:
   a. Requirements specification
   b. Simulation
   c. Hardware / software trade-offs
   d. Physical design rule checks

3. BE tools include:
   a. Synthesis
   b. Placement
   c. Design capture
   d. Simulation

4. Design views are:
   a. Top, bottom, and side views of the IC
   b. More expensive than designs without views
   c. Different descriptions of the design
   d. Different classes of tools

5. Design data
   a. Includes data content, format, and file arrangement
   b. Is provided by the tool vendor

c. Can be numbers, letters, test patterns, tool settings, etc.
d. Is one of the design views

6. Design hierarchy is:
   a. A design style
   b. Three classes of tools
   c. Physical layers on the IC
   d. A series of design views in order of increasing detail

7. Design styles include:
   a. Top-down, bottom-up, and middle-out
   b. Top-out, bottom-in, and middle-down
   c. Bottom-down, inside-out, and outside-in

8. Design partitioning
   a. Assigns functions to different blocks
   b. Assigns the design rules
   c. Separates the design from the data

9. Chip architecture refers to
   a. The levels of hierarchy of the chip
   b. The width, depth, and size of the chip
   c. The chip structure (such as a GA, SC, etc.)

10. Design methodology refers to
    a. The methods used to manufacture the chip
    b. The sequence of design steps used to develop the chip
    c. The kinds of tools used to make the chip

11. Design flow refers to
    a. The actual tools and tool steps used to design the chip
    b. The speed at which the design work flows
    c. The amount of work the engineers can do in one day

Answers: 1-a; 2-b; 3-b; 4-c; 5-c; 6-d; 7-a; 8-a; 9-c; 10-b; 11-a.

5 Electronic System-Level Design Tools

In this chapter...

INTRODUCTION

Electronic System-Level (ESL) design covers most of the whole electronic product. ESL includes the integrated circuits, power, and electrical and human interfaces.

A single IC can now contain most of the system electronics. Therefore, system-level design is often a part of the IC design. Software now resides on the IC as well, so we include embedded software as part of ESL design.

Let us join Nora as she learns about ESL design tools from Sam. He is a system engineer at SysComInc.

Nora: Thanks for helping me prepare for tomorrow's product meeting, Sam. Will one of our Sandbox tools help you design your planned product?

Sam: Yes, but actually it is our customer who is planning a new product. We are going to help their team define what they really want and need.

Nora: Don't they already know what they need?

Specification Guidelines

Sam: Yes and no—they have some ideas, but are unclear as to what can and cannot be done in a reasonable time and cost. There are usually several ways to solve a problem, and we can help them there. We follow guidelines to produce a useful requirements document. The requirements specification process is usually iterative.

Sometimes we discover constraints which no one mentioned or implied. These might be factors such as budget, time, management approval, existing environment, or user preferences.

Ideally, a knowledgeable customer, a knowledgeable marketing person, and a knowledgeable engineer can learn from each other. (Although that is admittedly an unlikely trio.)

There are some tools to help with the generation of requirements analyses and documentation.

Nora: So an early meeting is important. After you get the specification well-defined—what next?

SYSTEM-LEVEL DESIGN TOOLS

Sam: I think Andrea showed you a table of major design tools. Let me show you that table again, with the system-level tools highlighted. (See Table 5.1.)

Table 5.1 Major ESL Tools

ELECTRONIC SYSTEM-LEVEL (ESL) DESIGN	
Users	**Tools**
Architect	**ESL Design Entry**
System Engineer	**ESL Modeling**
IC System Engineer	**ESL Verification (Bench Test)**
	ESL Timing Analysis
FRONT-END (FE) DESIGN	
REGISTER LEVEL	
Users	Tools
	RTL Entry
System Engineer	Test Bench
Logic Designer	RTL Simulation
ASIC Designer	Formal Verification
Test Engineer	Design for Test
	Timing Design
	Thermal Design
	Power Design
	Signal Integrity Design
	Synthesis to Gates
GATE LEVEL	
Users	Tools
	Schematic Capture
Logic Designer	Gate-Level Simulation
ASIC Designer	
BACK-END (BE) DESIGN	
Users	Tools
	Floorplanning
	Layout—Place & Route
Layout Designer	Electrical Rules Check
Logic Designer	Physical Rules Check
ASIC Designer	Extractors & Delay Calculators
Test Engineer	Timing Analysis
	Power Analysis
	Thermal Analysis
	Other Analyses

Engineers use programming languages or graphical block design tools to enter design ideas into the computer. They use modeling tools to help make design trade-offs and try out ideas and different approaches to the problems. Test benches provide a test environment for easier design verification. Initial timing analysis helps predict the system performance.

High-Level Modeling

Sam: High-level *modeling* tools help estimate interactions between factors such as performance, power, weight, cost, and so forth. Problems such as traffic congestion on a highway or an inter-processor data bus are often modeled. Modeling helps ensure that the design can handle the traffic and avoid bottlenecks.

An architect makes sketches of different house plans to see how they meet the customer's needs. An interior decorator may experiment with several color combinations for a room. Similarly, the system designer may *model* parts of the design on the computer.

Did You Know?

Computer modeling is widely used to test ideas before building the actual bridge, building, airplane, spacecraft, or electronic system. Almost any characteristic or situation can be modeled: power, speed, traffic flow, or catastrophes.

Nora: What tools do they use?

Sam: Design architects usually model suspected problem areas using whatever software language they know best. Many use a variant of the general-purpose C language.

Other languages are useful for specific areas such as communications and multiprocessor systems. Usually, only the major unknown areas are modeled or estimated (e.g., power, performance, bottlenecks, or area).

Nora: They don't model the whole system?

System-Level Design Languages

Sam: Not usually, because the modeling effort would take too long. Sometimes, the whole system may be modeled at a very high level for interface or demonstration purposes. Often though, there are just a few areas which the engineers are not sure will work.

We need a single design language to describe the whole system, both hardware and software. Programmers want it to be compatible with the design software languages they use. Hardware designers want it to be compatible with the hardware design description languages they use. System architects want it to capture all the constraints of the system. They also want to be able to model the whole system or just parts.

Nora: So is there a common system description language now?

Sam: Not yet. Neither the basic C programming language nor a hardware description language has enough of the necessary features. Different new constructs have been added to both. These are competing forms and extensions of the C language and C++. Extensions of hardware description languages used for FE design (Chapter 6) are also competing. Several groups are trying to standardize one or more of these (complementary or competing) system description languages.

Nora: So a system-level language is coming, someday?

Sam: Yes, they are making progress.

Design Space Exploration and Trade-offs

Sam: There are always several approaches to addressing the project goals, with many conflicting or interacting constraints. System design evaluates these interactions. The system engineer must make trade-offs based on the evaluations.

Nora: Can you give me a few examples of trade-offs?

Sam: *Trade-off factors* may include size, cost, power, performance, development risk, beating the competition, or time to market. Many of these factors interact. The designer may trade off (evaluate) designing a slower IC in order to reduce the power for longer battery life. The goal is to achieve the best balance, satisfy all the requirements, and reduce the design risks.

Increasingly, the system functions are done with on-chip software. Therefore, a key decision is which functions to do in hardware and which in software. Implementation in hardware runs much faster but costs more and is harder to change. Modeling the design both as hardware and as software helps the system engineer make the decision. The result is a partition of the design into hardware functions and software functions.

Let me sketch you an example of the multilevel balancing of trade-offs. (See Figure 5.1.)

This shows only a few of the trade-off comparisons needed. For instance, I show cost against product volume and time-to-market. However, cost is also a trade-off with features such as speed and power. Furthermore, speed depends on power and chip size.

At another level, you have the hardware/software trade-off to make. This also impacts development time and time-to-market. Note also the trade-off decision between using in-house expertise and outside consulting.

Another trade-off decision involves what design pieces already exist that can be re-used and what portions must be a new design. Using existing portions of hardware or software already known to work reduces the design risk.

Did You Know?

The more standard parts one can use in a chip design, the easier and faster it is to implement. Chip designers can choose from large libraries of standard parts. They try to minimize the amount of custom design work, particularly on large chips.

Predesigned blocks of memory, processors, I/O controllers, and so forth are called *Intellectual Property (IP),* cores, macros, or virtual components (VCs). (These may be able to be dropped into a design as complete units.)

The high-level, fuzzy aspects include project costs, development time, market needs and risk. These all interact with detailed factors such as speed, power, size, and functional features. So trade-off decisions are more complex and intertwined than they first appear.

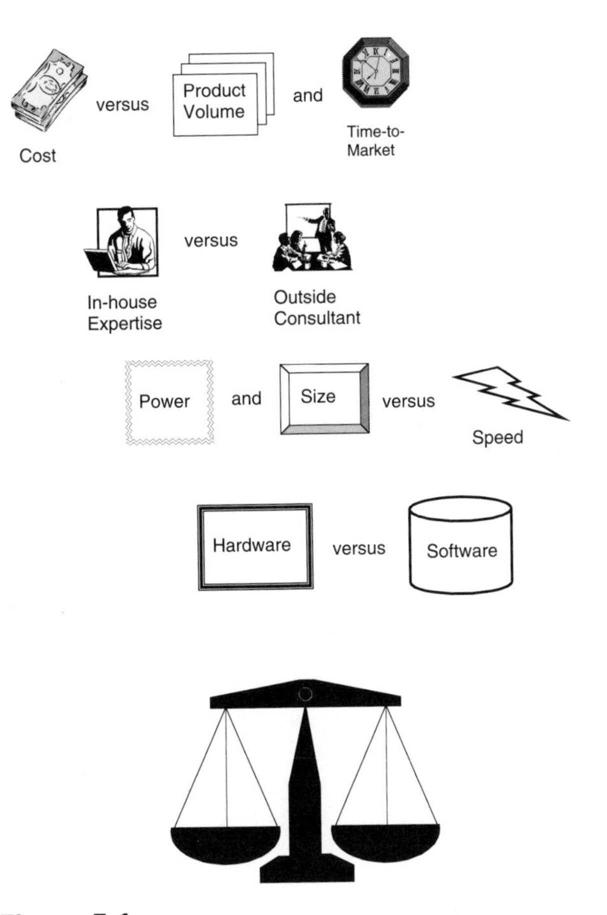

Figure 5.1
Design Trade-offs

In addition, we also have all kinds of constraints, legal restrictions, and design rules. For an electronic system or integrated circuit, the constraints include chip size, cost, and power. Others are clock speed (performance), and TTM. Sometimes they are rigid constraints, sometimes only preferences. The constraints form a series of fences or boundaries. We use EDA tools to work within the area or *design space* defined by the constraint boundaries.

A system engineer may model several trial designs to make intelligent trade-offs within the design space. What happens to the chip power and cost if we design for performance? Is 25% lower power more important for the application than 10% higher speed? This what-if process is known as *design exploration*, and may affect the final specification.

Nora: I see that a lot of hard choices get made, Sam. How do the engineers know if their final design choices will work?

Test Bench Creation

Sam: That's an excellent question. The ESL design team must develop system-level tests which verify that the design works as **designed**. The tools must also validate that the system does what the **customer expected**.

It takes a lot of work to develop an environment for testing a system design. *Test bench creation* (automatic or semi-automatic) is another kind of system-level tool.

A test bench is a model of the external world within which the system (or subsystem or chip) is used. It can generate the inputs to stimulate the system under test, and measure or display the outputs of the system. It assists the designer in verifying and validating the system design.

Testing or *troubleshooting* identifies the causes of design or test problems. Removing or fixing the problems or *bugs* is called *debugging.* And retesting after fixing a bug is important to ensure that the fix did not create a new problem.

Nora: So an EDA tool helps create a test bench?

Sam: Yes, the tool helps both create and operate it. Test bench tools are getting more intelligent all the time. The earlier a problem is identified in the design process, the lower the cost to go back and fix it.

The data volume, tool cost, and people cost are highest in the back-end. Most labor (and design problems and bug-fixing) historically takes place in the back-end design.

Did You Know?

Verification and validation of chips is a very difficult and time-consuming process. A chip may have hundreds of inputs and outputs and perform dozens of operations. It could take many years to test all possible sequences. For example, a computer chip must handle all kinds of data (numbers, text, pixels, etc.) It must handle many possible errors (caused by noise, failed connections, misroutings, and the like).

Just adding a long series of numbers in every conceivable order could take hours. In practice, one cannot test everything. The designer must select a reasonable set of test cases that covers all the common situations. However, it must not take a ridiculously long time to check. Usually, the designers check only typical values and boundary cases (values at the minimum or maximum points).

Nora: I've even heard about bugs on the news, especially when they involve Intel or Microsoft.

Sam: Yes, even large companies cannot check everything. Reporting on bugs can be worldwide news items, not just a local embarrassment.

Nora: You mentioned predicting chip performance. How do you do that?

Sam: We do that using *static* and *dynamic timing analyzers*. These EDA tools are used at several levels in the design sequence. Static analyzers add up all the block, gate, and wire delays to uncover obvious timing problems.

Dynamic analyzers calculate delays while simulating the actual chip operation. Both help identify potential timing bottlenecks. Engineers need to know about timing problems as early as possible to ensure system performance (speed).

Nora: What other system-level tools are there?

Other System-Level Tools

Sam: We have many complementary EDA tools to assist the system engineer. A few of them are:

- Editors, compilers, debuggers, and other tools for high-level design languages
- System-level timing and throughput analysis tools

- Special tools for specific applications such as *Digital Signal Processing (DSP)* design, communications system design, embedded system design, or multimedia chip design
- Special tools for FPGA design and other specific hardware implementations
- Analog, mixed-signal, and RF design tools

Nora: What are the last three?

Sam: Wireless products use *radio frequency (RF)* circuits. Analog circuits work with continuously varying signals found in the *real world* (non-digital). Analog signals may be from voice, temperature, pressure, sound, or light sensors. Mixed-signal circuits convert analog signals to digital and vice versa. (See also Appendix C.)

Did You Know?

Analog portions of a system IC are typically only 5% of the total chip. However, they are often the limiting factor in the system performance.

HARDWARE/SOFTWARE INTEGRATION ....

Nora: That's quite a few kinds of system-level tools. I heard that hardware / software integration was also a major problem. Is that true?

Sam: Very true, I'm afraid. Normally, the hardware computer or special processor is designed first. Later, application software is integrated and tested on the new hardware. This sequential process is the *hardware / software integration.* Let me sketch you a quick picture of the integration process. (See Figure 5.2.)

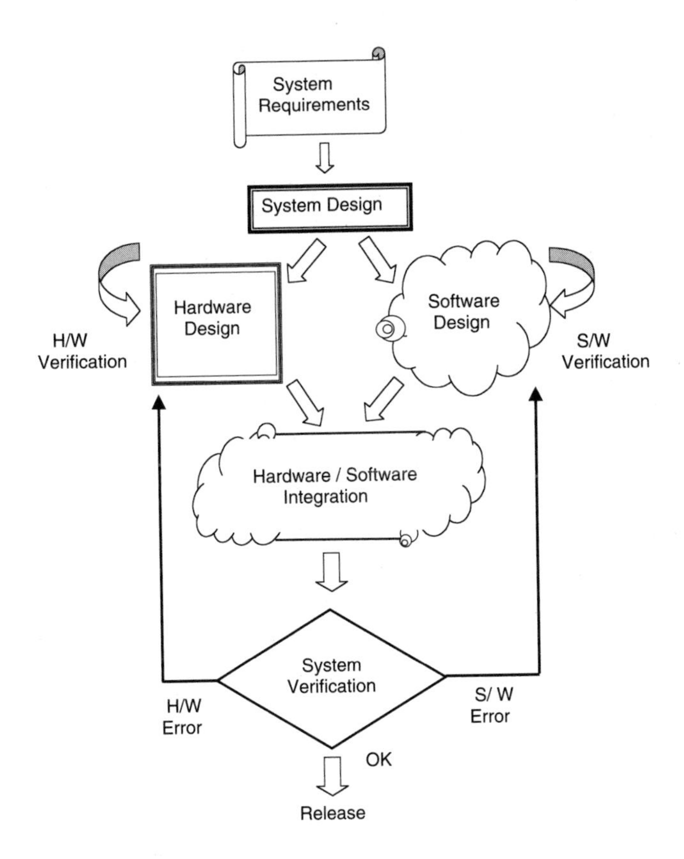

Figure 5.2
Hardware/Software Integration

Note that a soft cloud represents the *software*, and a hard box shows the *hardware*. The integration is the combining of these two.

As we discussed earlier, the system designer partitions the functions into hardware or software. These functions are designed in more detail and are then integrated.

The integration, of course, suffers from any errors made in both hardware and software development. But the problems often show up during the integration verification process. Since the hardware is more difficult and costly to change, most effort tries to fix all problems in the software.

Nora: Is there a solution to the integration problem?

APPROACHES TO CO-DESIGN ....

Hardware and Software Co-Design

Sam: A partial solution. People have looked for ways to do software development *concurrently* (in parallel) with the hardware design. The goal is *hardware/software co-design*, which means that both can proceed at the same time. We could fix problems uncovered by either group, earlier and where most appropriate.

Nora: Is there a way to do that?

Sam: We have several approaches to provide an early hardware model for software development. These include hardware simulation, hardware emulation, rapid prototyping, and virtual prototyping.

Hardware/software integration and *co-verification* can be done incrementally, if a hardware model can be generated quickly. Then the early software can start testing on the hardware simulator, emulator, or virtual test bench. The "co-" part of the term refers to verifying both the hardware and software designs concurrently.

Hardware simulation uses software to model individual transistors, gates, and wires. It can check the hardware functions, but it runs thousands of times slower than the actual hardware.

Hardware emulation involves configuring PC boards full of FPGA hardware chips or processors and existing chips. The result is a hardware model that is fast and easy to implement. However, it is much larger and slower than the final chip implementation. It is useful to debug the logic and provides an early platform for software development. Several EDA companies sell hardware emulation systems.

The FPGA and interconnect time delays are also vastly different from that of the final hardware. Therefore, the hardware timing is not accurate with the FPGA emulation approach.

Rapid prototyping uses both FPGAs and software on existing processors to emulate other system pieces. It matches system interactions in function and timing better than either pure simulation or emulation.

Rapid prototyping works well where major pieces of the total system are known or already exist. It helps the incremental integration of the many blocks. Prototyping has long been used and is sometimes custom-made. A few vendors offer customizable modular systems to speed up and simplify the building of "rapid" prototypes.

An emerging version of rapid prototyping is the *virtual prototype*. This EDA framework can run software or modeled hardware or both. Mixed levels of hardware and software can be prototyped together (e.g., high-level software models mixed with low-level hardware blocks). This helps start integration at earlier design stages.

Nora: Now I'm sorry I asked. It must really be a problem if there are so many solutions.

Sam: Yes, and I'm sorry for the long answer—but you did ask. Incidentally, your company is working on one of these solutions. We can summarize these approaches in a table. (See Table 5.2.)

Table 5.2 Co-Design Approaches

NAME	APPROACH	COMMENTS
H/W simulation	H/W simulated in software	Very slow
H/W emulation	FPGA hardware racks	Fast for functional verification
Rapid prototype	Semi-custom FPGA H/W	H/W prototype
	Microprocessors for S/W	Function & speed
Virtual prototype	All software prototype	Can run H/W or S/W model

Nora: Thanks, Sam. I see that you have to deal with both hardware and software. I've also heard "embedded systems" mentioned. What are they?

EMBEDDED SYSTEMS

Sam: Traditionally, they were applications with a computer embedded in the system but invisible to the user. Examples are car engine controllers or cell phones. *Embedded system* now refers to any system with fixed software on the chip. These systems include both real-time, critical applications, and non-critical, non-real-time applications.

As processor speeds have steadily increased, more parts of the job can be done in software. If fixes are needed, software has the advantage of easy modification, compared to expensive IC changes. It is also easier to add features and upgrades in software. Embedded software (resident on the IC) is frequently used to achieve quick TTM.

Nora: What do you mean by "real-time"?

Real Time

Sam: Embedded systems include those critical industrial applications requiring high reliability and fast response in *real-time*. A simple example of a real-time system is a missile. The missile controller electronics must give the proper guidance signals at the proper time. Guidance directions (no matter how accurate) are of no use after the missile already has hit the ground.

Reliability

Sam: Many industrial applications such as power plants or oil refineries require high reliability. Having the computer system go down occasionally may be tolerable on a personal computer. It is annoying but not critical. However, that sort of crash is just not acceptable on critical industrial systems.

Did You Know?

Most users know that a personal computer occasionally suffers a catastrophic error. It then has to be reset (or rebooted, as the computer people put it). This is merely annoying with a desktop or laptop system. However, imagine the same thing happening to your car engine computer while you were speeding down the road!

To ensure software reliability, the industrial embedded systems industry typically uses different operating systems and design disciplines from personal computer software design. EDA tools now have to ensure the reliable design of both hardware and software.

Nora: Thanks, Sam.

SUMMARY

The system design constraints (rules, specifications, and guidelines) define the design space. System engineers use computer modeling to make design trade-offs within the design space. They explore trade-offs such as cost, performance, development risk, or TTM.

Re-use of standard known parts can speed the design and reduce the risk. These parts are called intellectual property (IP) and include blocks such as microprocessors, memory, encryption modules, and I/O interfaces. (Re-use is often difficult to achieve in practice, however. Appendix F discusses re-use and IP in more detail.)

System tools include high-level modeling programs, system-level description languages, and automatic test bench generators.

Historically, the hardware is designed first and then the software is integrated and tested, using the hardware. Errors in either or both can cause schedule slips. Approaches to hardware/software co-design and integration include simulation, emulation, rapid prototyping, and virtual prototyping. All of these use EDA support software tools, which are adapting to verify both hardware and software.

Hardware processors are often embedded on the chip. Operating system and application software is increasingly embedded on the chip as well. Real-time systems required reliable software and tight control of the chip timing. Embedded software now refers to any software on the chip, real-time or not. Therefore, system-level EDA tools now need to model software timing and reliability aspects.

QUICK QUIZ

1. System-level design starts with:
   a. Customer requirements
   b. Simulation
   c. Software development

2. System-level design tools include:
   a. Place and route
   b. Gate-level simulation
   c. Modeling

3. System-level modeling does **not** help predict:
   a. System cost
   b. System performance
   c. Customer requirements

4. System-level design languages usually:
   a. Describe low-level design details
   b. Model unknown or problem areas of the design
   c. Model only the hardware portion of the design

5. System design trade-offs do **not** include:
   a. Features, constraints, or preferences
   b. Cost, performance, or power
   c. Travel time, business meetings, or sales commissions

6. A document of requirements and design trade-off decisions is a:
   a. Project book
   b. System specification
   c. Test plan

7. The design approach of using existing blocks of hardware or software is called:
   a. Re-use
   b. Test bench
   c. Iteration

8. Design exploration evaluates many of the:
   a. Tests
   b. Trade-offs
   c. Specifications

9. EDA test bench creation tools help:
   a. Create a system environment to evaluate the chip design
   b. Build a system chip specification
   c. Decide on hardware re-use

10. System design tests usually:
   a. Cover all possible situations and sequences
   b. Cover typical cases and sequences
   c. Test only the hardware portions of the system

Answers: 1-a; 2-c; 3-c; 4-b; 5-c; 6-b; 7-a; 8-b; 9-a; 10-b.

6 Front-end Design Tools

In this chapter...

- Introduction
- Design Capture Tools
- Verification Tools
- Timing Analysis Tools
- Design for Test Tools
- Power-Related Tools
- Synthesis Tools
- Summary
- Quick Quiz

INTRODUCTION

The front-end EDA tools help develop the IC design and verify its correctness. EDA tools have replaced many of the most tedious manual design tasks. Table 6.1 shows the major EDA programs with the FE tools highlighted.

Other tools for areas such as signal integrity and design for manufacturing are used in both the front-end and back-end. We discuss those in the chapter on back-end tools.

Let us join Nora of Sandbox, Inc., as she talks with Luigi, a logic designer at Federal Semiconductor.

Nora: Thanks for meeting me for lunch, Luigi. Sandbox has many tools, so I am trying to understand what they all do.

Luigi: Glad to help, Nora. You said you wanted to discuss the front-end tools. Why don't we start with design capture?

DESIGN CAPTURE TOOLS

Luigi: *Design entry* or *design capture* tools allow the engineer to enter design ideas into the computer. Entry is done using either graphical symbols or textual descriptions. The design can be displayed in several forms or views.

Different tools work at different levels of design abstraction (or detail), so you need to understand what those are. Let me describe each for you. I'll explain the jargon as we go.

Device modeling is the lowest level of abstraction and is used by physicists and IC manufacturers to model transistor devices. Most chip design starts at the next level of abstraction, circuit design.

Circuit design uses diagrams with graphical symbols to represent devices such as transistors, resistors, or capacitors. Lines between device inputs and outputs represent interconnect wires. Each symbol corresponds to a software model of the device on the computer. This graphical view is called a *schematic diagram*. A textual description view can also be created using a circuit description language.

Table 6.1 Major Front-end Tools

ELECTRONIC SYSTEM-LEVEL (ESL) DESIGN	
Users	Tools
Architect	ESL Design Entry
System Engineer	ESL Verification
IC System Engineer	ESL Modeling
	ESL Timing Analysis
FRONT-END (FE) DESIGN	
REGISTER LEVEL	
Users	**Tools**
	RTL Entry
System Engineer	**Test Bench**
Logic Designer	**RTL Simulation**
ASIC Designer	**Formal Verification**
Test Engineer	**Design for Test**
	Timing Design
	Thermal Design
	Power Design
	Signal Integrity Design
	Synthesis to Gates
GATE LEVEL	
Users	**Tools**
	Schematic Capture
Logic Designer	**Gate-Level Simulation**
ASIC Designer	
BACK-END (BE) DESIGN	
Users	Tools
	Floorplanning
	Layout—Place & Route
Layout Designer	Electrical Rules Check
Logic Designer	Physical Rules Check
ASIC Designer	Extractors & Delay Calculators
Test Engineer	Timing Analysis
	Power Analysis
	Thermal Analysis
	Other Analyses

Logic design is the next level above circuit design. (Several transistors connected together can make a logical switch or *gate*). There are a variety of gate types used to make complex logical decisions. The logic designer normally works with logic gates chosen from a list or library of available, predesigned gate types. Logic design can be done using either a logic schematic symbol view or a textual entry view.

Nora: Yes, I know a little about the logic gates.

Luigi: Let me draw you some quick examples of a transistor circuit and logic schematics. (See Figure 6.1.) Note that schematic entry may be at the circuit level or the logic gate level.

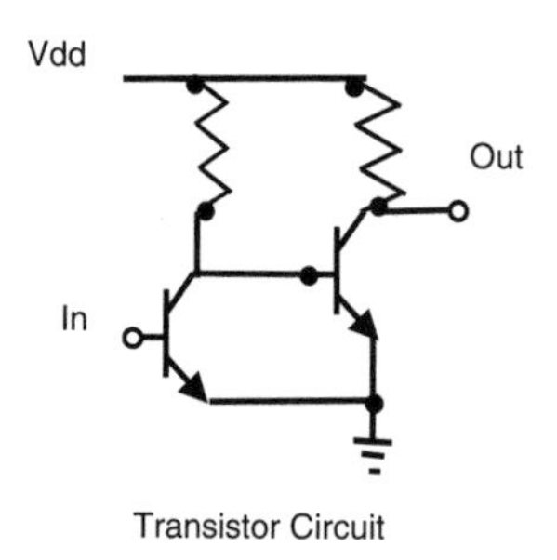

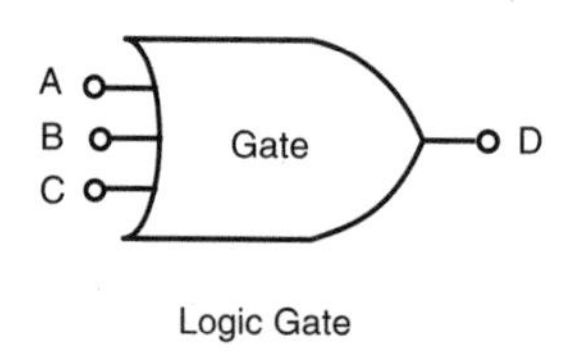

Figure 6.1
Schematic Examples

Luigi: The logic designer's library also includes *flip-flops*. Each flip-flop can store one bit of information (i.e., a one or a zero).

A row of flip-flops forms a *register,* which can store a *byte* (eight bits) or more of data. Data *words* may be anywhere from two to eight bytes long. *Memory blocks* can store thousands—even millions—of words of data.

Did You Know?

How big is a thousand? A million? Big numbers can be hard to grasp without a visual example.

Most of us have seen little black poppy seeds on rolls, bagels, and loaves of bread. They are about 1/32 of an inch (or about one mm) in diameter.

If you put these small seeds in a row across a regular card table (~ about one yard or meter square), it would take about 1,000 seeds. If you covered the whole table surface, you would have used one million (1,000,000) seeds.

If you covered the table four inches deep, you would have used about 100 million seeds. That's about how many transistors can fit on an IC chip the size of your fingernail!

Nora: With that many transistors to deal with, I see why we need EDA tools.

Luigi: Yes, it would take us years to do what the tools can do in minutes. Moving up a level from gates, logic designers do *Register Transfer Level (RTL) design.* RTL design consists largely of three actions:

- Moving data between registers and memory
- Manipulating the data (e.g., adding or comparing)
- Storing the result in another register or memory location

The Register Transfer Level (RTL) design information is entered using a *hardware description language (HDL).*

Hardware Description Languages

Luigi: Hardware description languages have largely replaced the schematic entry approach for logic design.

The HDL can handle several levels of abstraction, including *behavioral, register transfer,* and *gate levels.* (*Behavioral design* describes **how** the data is manipulated.) The HDL describes both *movement of data* between registers (register transfer) and the gate-level *control logic.*

Nora: What is control logic?

Luigi: Control logic describes how or when something gets done. You are familiar with controllers and control logic in daily life. For instance, a thermostat controller turns the furnace on and off in a house to control the temperature. A lamp controller turns on a light at a certain time or when it gets dark.

Nora: Okay, so the designer has to enter all the control and data information into the computer?

Luigi: Yes, that is design capture.

Nora: Is the hardware description language in English?

Luigi: No, unfortunately, human languages are too complex. HDLs are simpler.

Did You Know?

A hardware description language is just a programming language suited for describing electronic circuits. There are many similar specialized languages for other fields.

English is not used as an HDL because it is too ambiguous. HDLs are defined very formally in syntax (symbols and structure) and in semantics (meanings and relationships). They are much simpler than English, with typically only a few hundred words and rules.

Luigi: Two standard HDLs emerged—Verilog (originally a vendor HDL) and VHDL (a standards group HDL). The Accellera standards group now manages both languages, so both are open and available. EDA software has been developed for both languages. Eventually, VHDL and Verilog will merge or one will supplant the other. Some tools can translate from one to the other for interactive editing in either representation. (We also have tools that can convert a schematic to an HDL and vice versa.)

Did You Know?

VHDL came out of a U.S. Department of Defense program called VHSIC (Very High-Speed Integrated Circuits). The military/aerospace IC design market adopted it, as did most of Europe, while Verilog became the de facto U.S. commercial standard.

Luigi: Let me show you a simplified HDL-like description. (See Figure 6.2.)

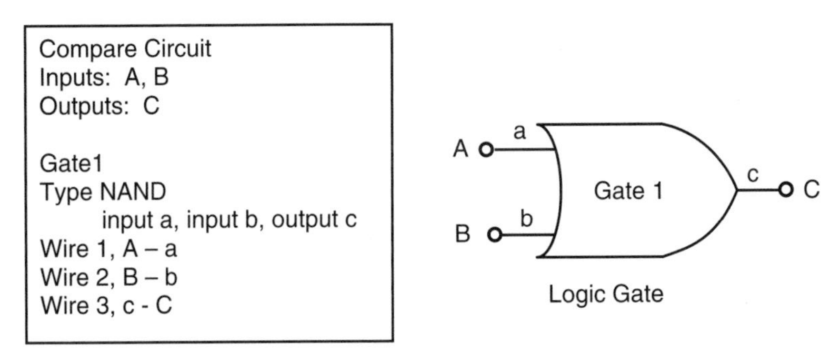

Figure 6.2
Hardware Description Language

Real HDL code is a bit more technical. Note that the text is a description of the graphic gate schematic next to it on the figure.

Luigi: Moving up yet another level, design ideas are often described with *block diagrams.* Here is a block diagram example. (See Figure 6.3.)

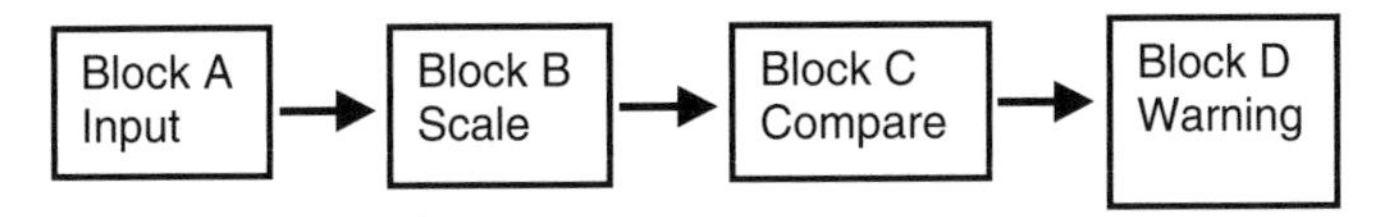

Figure 6.3
Block Diagram Description

The block diagrams show the functions needed and their sequence or relationships. The implementation method for each block or function (such as H/W or S/W) may be still undecided. Modeling languages are sometimes used to describe functions and bus data transfers at this level.

Nora: So which entry method is better?

Luigi: It's a matter of experience and preference. All methods are used at different stages of design. Due to increasing chip complexity, designers are using higher-level languages, and a computer generates the lower-level detailed descriptions.

Specialized Design Tools

Nora: Are those the only methods to enter the design?

Luigi: No, there are tools for designing specific subsystems. These include memory systems, network interfaces, and peripheral interfaces. Some design tools focus on specific building blocks such as processors or memory blocks.

However, in whatever way the information is entered, the result comes out the same.

Nora: What is the result?

Netlist Output

Luigi: The final output of the HDL or schematic is the design netlist. This is a listing of all the gates, flip-flops, registers, and their interconnections (nets). The netlist is compiled from the HDL or a synthesis tool.

Synthesis is like turning an architect's drawings into a parts list which a builder can use. You go from general descriptions of wood, cables, windows, and so forth, to actual part brands and model numbers.

For a system-on-chip IC design, the front-end may include ESL design tools. As you know, there are several efforts to standardize an abstraction-level language above RTL. The goal is to allow direct conversion from that language into either an HDL description or into a software language description.

Nora: Yes, Sam discussed that with me, and he mentioned the many chances for design errors. Isn't the HDL design entry prone to errors?

Design Capture Checking Tools

Luigi: Sure, and that's why we have some checking tools. After we capture the hardware design, we check the HDL code for grammar and syntax errors. The EDA tools are called *HDL checkers*, and they work similarly to a word processor spelling checker.

Did You Know?

An early checker for C programs was called "lint" (lexical interpreter). The checking program would clean up the syntax errors. They would "get the lint out of " (or "de-lint") the software code. The "lint" checker term later got applied to a variety of HDL check programs.

VERIFICATION TOOLS

Nora: After the design description is entered and de-linted, then what?

Luigi: Once a design description is "clean," we need to verify its functions (does it work as expected?). The designer prepares a sequence of inputs to the logic (block or chip) with expected outputs.

The inputs consist of a set of signals going into the logic. A set of output signals comes out of the logic. The digital signals are two level (high or low) electrical voltages. You recall that the two voltages can be called ones and zeros as well.

These input and output data sets are known as *test patterns* or *test vectors*. They are prepared manually, or with the help of an EDA test bench tool. A test bench is a software platform that helps to create test patterns and to run the tests.

We can talk later about other EDA tools which automatically generate additional vectors for manufacturing test.

Design Verification

Nora: Why is it so hard to verify the design?

Luigi: Errors occur in the requirements specification, the engineer's design ideas, the design capture, and in the tests themselves. That's why the verification consumes so much of the development time.

Nora: I think someone told me that it takes 60-70% of the design effort.

Luigi: That's probably about right. As complexity increases, the verification time also increases. Let me give you a quick overview first. Modeling tools allow the designer to model key functions at a high level of abstraction. Design decisions such as choosing to do a function in hardware or software are usually modeled and evaluated at the electronic system design level.

Simulation tools verify design behavior (e.g., what it does) or performance (e.g., how fast it runs). Software simulation is used widely to try things that would be too expensive or dangerous to do directly. Aircraft simulators can train pilots without risking an actual aircraft. Power simulation is important for battery life estimates on laptops or cell phones.

Formal Verification tools rely on notes (*assertions*) which the designer adds into the design or as a separate file. These notes describe the *intent* of what the logic should do (*properties*). The tools check that the design is consistent with these properties.

Some tools use formal mathematical approaches to prove that a design is self-consistent or logically correct (like a grammar checker). Others check that the logic remains *equivalent* before and after a transformation from one level of abstraction to another. (However, just because the design is consistent doesn't guarantee it will work as expected.)

Analysis and checking tools perform all sorts of detailed error and rule checking on the design, including function, logic, circuit, timing, and power usage.

Other checks include preventing problems with layout, thermal design, power distribution and manufacturing constraints. Some checks are done at the front end, some at the back end, and some at both.

Nora: So, there's a whole lot of checking going on. Can you explain a little more about EDA simulation?

Simulation

Luigi: Sure. The simulation is the actual running of the tests I mentioned earlier. The design input to a simulator program **is** the HDL description (Verilog or VHDL). So we talk about "running the Verilog simulator."

An EDA *simulator* stores the design netlist and then applies a set of input test patterns to the logic. The designer (or test generator program) creates test patterns and expected outputs.

The simulator traces through the logic, finding the logic values (ones and zeros) of each gate in turn. The simulator follows all the logic paths to the outputs and compares those to the expected outputs (listed with the input test patterns). This uncovers functional, logic, and connection errors.

Nora: So the simulators are driven by the test vectors and check whether the design works correctly?

Luigi: That's right. In addition, they can also check for timing errors. Simulators use gate, register, clock, and interconnect *delay models* to calculate timing through the logic block. There are several kinds of simulation.

In *event-based simulation*, the simulator keeps track of the switching of each gate (an *event*). *Clock-based simulation* keeps track of changes only at clock times.

For a computer design, *cycle-based simulation* updates values at each computer instruction cycle. It is *cycle-accurate* if the timing is also correct within the instruction cycle.

Luigi: Let me summarize verification with a quick sketch. (See Figure 6.4.)

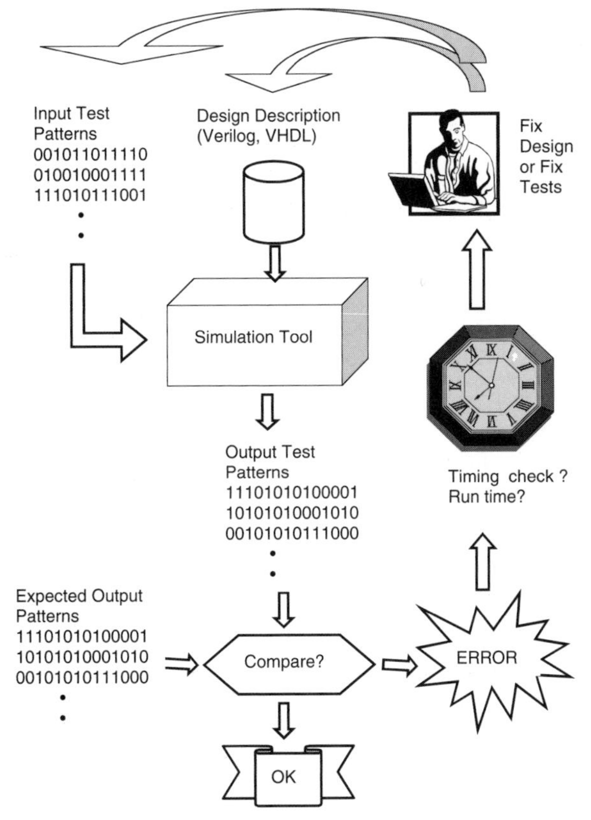

Figure 6.4
Verification

Luigi: Note the iteration loop: design, test, compare, fix design, or fix test, and repeat. Also note the clock—both the design timing and the speed of the simulation (*run time*—how long the simulation takes) are important.

Simulation Speed

Nora: I understand simulation can take a long time.

Luigi: Yes, the simulators can take many, many hours to run large designs. Any time we find and fix a problem, the simulation must usually be completely re-run. Various approaches have been made to speed up simulation. A few vendors make special computers optimized to accelerate the running of simulation programs. Some *simulation accelerators* can run simulation a thousand times faster than software alone.

Nora: Sam, the system engineer, told me about the hardware emulators.

Luigi: Yes, and other companies use multiprocessors working in parallel to speed simulation. Other approaches use a roomful of loosely coupled standard processors to speed up simulation.

Nora: Those must be the “server farms” that Hugo, the design manager, mentioned.

Formal Verification Tools

Nora: Okay, you can speed it up, but how do you know you have got a thorough test?

Luigi: That’s an excellent point. For simple control logic designs, we can predict all possible (*exhaustive*) tests. For complex designs such as computers, however, we cannot predict all possible sequences of instructions.

A *formal analysis* tool can check that the logic meets much of the designer *intent.* But it still cannot guarantee that the design will actually work. In practice, we usually need several kinds of verification.

Remember we have a general netlist before synthesis and one with specific parts after synthesis. The synthesis program might change the logic. *Equivalence-checking* tools evaluate the netlists before and after synthesis to ensure that they are logically the same. This is another kind of formal verification.

Device and Circuit Simulators

Nora: Are there other kinds of simulators?

Luigi: Yes. There is a whole other class of simulators for semiconductor devices and transistor circuits. *Device models* and *device simulators* predict the electrical operation of the *semiconductor devices* from the underlying semiconductor physics.

Circuit simulation programs predict the electrical circuit operation using mathematical models for all the components and wires.

Did You Know?

One of the oldest circuit simulators is SPICE (Simulation Program for Integrated Circuits Emphasis—University of California at Berkeley, 1975). Many variations exist, but it is still an industry standard for predicting the operation and time delays in a circuit.

TIMING ANALYSIS TOOLS

Nora: Is there any shorter way than simulation to get the timing right?

Luigi: Yes, there are two kinds of timing analyzer tools—dynamic and static timing analyzers.

Dynamic Timing Analysis

The simulator can use known or estimated values for the wire and gate delays. It calculates the event-to-event times while the simulation runs. This is *dynamic timing analysis (DTA).* It can reveal potential timing errors. A timing error example is data arriving late at a register, after the clock signal.

Static Timing Analysis

In addition to simulation, there are *static timing analysis (STA)* tools. These move a value through the logic and uncover timing errors without using test vectors. These analyzers run much faster than simulators.

Clocks

Nora: You've mentioned clocks a few times. How do those fit in?

Luigi: Timing is a critical design area. High-speed electrical pulses called *clocks* control most digital systems.

Did You Know?

Most digital systems use clocks (timing pulses) to keep everything in step. These are not wall clocks.

Everyone has seen marching bands on TV, at sports events, and in parades. Each member of the band moves in time (*synchronized*) with the drumbeat. If the beat rate is faster, they move faster.

Everyone has seen heartbeat monitors on TV drama shows about hospitals and doctors. The screen shows the electrical pulses of the heart beating (until it flat-lines and the show is over.) Digital clock pulses look very similar on a monitor.

The more frequent the drumbeat or heartbeat, the higher the frequency (beats or pulses per minute). Digital systems measure frequency in millions of clock pulses per second.

A 100-megahertz chip has a top clock rate of 100 million clock pulses per second.

Clocks act like traffic lights controlling when cars (signals) move, and when they wait for others. Fire trucks or ambulances (and some signals) are more critical than others.

Clock signals need to arrive at thousands of places all over the chip at the same time. This often requires moving wires around or shortening wire paths. The IC performance depends on the wiring layout of these clocks.

Nora: What happens if a signal is late?

Signal Timing

Luigi: A *race condition* (or *delay fault*) occurs if a data signal arrives too early or too late with respect to the clock. The signal may not get latched into a flip-flop or register if it is late.

Signals also get delayed as they go through gates and wires. Signals may arrive earlier or later than expected (*slack time*). So slack times are calculated for every signal net.

Timing analyzer tools search the design for race conditions. They identify potential faults in the FE design. They identify (and sometimes fix) real faults in the BE physical design. The slack time can be changed by changing the wire lengths or inserting a *buffer circuit* which speeds up the signal.

Modifying the design for thousands of signals so that none arrives too early or late is known as *timing closure.* However, the fixing of one timing problem may cause another problem to pop up somewhere else. This can make timing closure very hard (or impossible) to achieve on large chips. One way to achieve closure is to slow down the clock rate. Some integrated tool suites do this to guarantee closure.

Nora: But then you may not meet the IC performance requirement, right?

Luigi: That's right.

Nora: It seems there are tests at every level of design entry, modeling, simulation, and so on. So verification must be a really important part of design.

DESIGN FOR TEST TOOLS

Luigi: Absolutely. But there is a difference between the test for design verification and *manufacturing test.*

In design verification, we are trying to verify that the actual logical design works and does what we intended it to do. We do this both statically (without timing) and dynamically (with timing). It is less expensive to fix a design error the earlier it is caught. So we try to do as much verification as possible, and as early in the design process as possible.

In manufacturing test, we assume that the design is correct and focus on the manufactured parts. A defect in manufacture may cause a chip to not work. Many tests must be run on each part and each test adds to the chip cost. So we want to use as few tests as possible and to screen out defective chips as early as possible in the manufacturing and assembly flow.

Design for Test

Luigi: The designer needs to plan for testing the chip early in the design. Several design test approaches are collectively known as *Design For Test (DFT).*

Testing quickly gets very complex and time-consuming. For instance, look at all the functions on a cell phone—messaging, telephone number book, ring patterns, displays, etc. Imagine having to test all these functions, and in different sequences. There are several test tools to help the designer generate, execute, and document test plans.

Did You Know?

Two things are needed to do thorough testing. One is a means to access the all the logic inputs. The other thing is a means to observe all the logic outputs.

Testing a chip is a little like a spy looking into a building through keyholes. The keyholes are the few signals coming off the chip. You can't easily get to see everything inside. So internal logic *test points* are designed in and brought off the chip. They ensure both access to and observation of the logic. Boundary scan logic and boundary scan registers are standard ways of doing this.

Boundary Scan

Luigi: *Boundary scan* is one Design for Test (DFT) approach. With large chips, registers are inserted at the edges or boundaries of logic blocks. Test vectors can be *scanned* in and out of these registers to test sections of the whole chip. EDA tools can insert these registers automatically. (This is implemented in the BE design phase, but I mention it here as part of the overall test planning done in the FE phase.)

Built-in Self Test

Luigi: Another DFT technique is called *Built-in Self Test (BIST)*. This approach generates its own patterns **on-chip**. It does not involve moving test patterns in and out of the chip, so it runs much faster than boundary scan.

BIST can be used for testing both logic and memories. Several EDA tools automatically insert standard BIST registers and logic into a design. One drawback is that the on-chip test pattern generator does not always guarantee to cover all possible faults.

Nora: Okay, you test function and timing before implementation, after implementation, and at manufacturing. So is that all of the tools?

Luigi: I'm afraid not—there's more.

Nora: There seems to be no end to design problems and tools.

POWER-RELATED TOOLS

Luigi: Yes, there are many. And more problems arise as the semiconductor technology keeps shrinking. Power is a good example.

Power Estimation Tools

Luigi: Front-end designers use power calculation tools to estimate the power. Too much power will overheat the chip—we need to be sure we stay within the power budget early in the design. (Power is checked again in the back-end.)

Also, power is a major trade-off factor with speed—faster transistors tend to use more power.

Nora: How about portable things such as my cell phone?

Low-Power Design Tools

Luigi: Portable personal electronics is a growing area. Battery life is critical in most of these products and is a major selling feature. Therefore, low power is now a primary design focus.

Control of the consumed power is achieved by many different design techniques. We have approaches at all levels of design (system, logic, circuit, transistor, and process).

There are EDA tools to help achieve low-power designs. For example, they reduce clock frequencies and identify when logic blocks can be turned off.

Nora: How do we get from the front-end tools to the back-end tools?

SYNTHESIS TOOLS

Luigi: With *synthesis,* which our last tool for discussion.

Nora: I've heard synthesis mentioned, but I didn't quite understand it.

Luigi: Synthesis is the bridge between abstract logical design and the concrete physical design. Most front-end design views are descriptions or graphic symbols of gates, flip-flops, registers, and memory blocks. (These are *symbolic* design views with little real *physical* design information.)

However, real hardware is built with mask patterns used by the semiconductor manufacturer. These physical views describe the devices as they look on the masks and actual chip. The devices are the transistors, wires, resistors, and so forth. Their physical details such as length, height, and area are described in a *device library*. The device libraries are related to specific semiconductor processes (e.g., 0.15 micron or 0.13 micron).

For each device in the symbolic design, the designer must select a physical device from a device library. This tedious, error-prone mapping process is largely automated by EDA synthesis tools.

The programs do some optimization for power, speed, or area. Some synthesis tools include physical layout and wire length optimization issues as well.

Automatic synthesis is also useful for *design re-use*. An existing design or IP block can be *re-targeted* to a faster, denser process. Since new processes are developed about every 18 months, automatic mapping of an existing chip or IP block design to a new process device library can be very useful.

The output of the synthesis tool is a netlist similar to the one used in the FE design. However, the devices in this netlist have real physical and timing views.

Luigi: That's about it for the front-end tools.

Nora: Thanks very much, Luigi.

SUMMARY

FE chip designers enter use design capture tools to enter their logic design ideas into the computer. These tools support block diagrams, schematic diagrams, or hardware description languages. The output is a netlist of logic gates and interconnecting wires.

Simulator tools verify the design using the netlist and test patterns developed by the design engineer. Test bench tools assist the test generation work. There are several kinds of simulators and simulation accelerators. As chips grow more complex, it takes more compute power to simulate them in a reasonable time.

Formal verification can catch some design errors by checking for the design intent. Static timing analyzer tools can check for timing errors more quickly than simulation.

Digital systems use high-speed clock pulses to synchronize the logic functions. Timing faults can occur if signals arrive at flip-flops or registers after the clock has occurred. Timing closure is the critical process of fixing all the timing errors. Hopefully, the chip performance goals can be met as well.

Manufacturing tests supplant the engineering tests after a first chip is proven to work. Automatic test pattern generation creates optimized test patterns to check all stuck-at faults. Boundary scan and built-in self-test can speed up the chip testing.

Low-power design tools and techniques are used in portable applications. Power estimation tools ensure that the chip will not get too hot or exceed the power budget.

Synthesis is the bridge from FE to BE design. It maps the FE netlist to a process-specific netlist with physical views.

QUICK QUIZ

1. Design entry gets design ideas from the designer into the
   a. Computer
   b. Library
   c. BIST

2. Logic design involves
   a. Resistors
   b. Gates
   c. IC packages

3. An example of a hardware description language is
   a. Verilog
   b. STA
   c. EDA

4. The output file from the FE design is called:
   a. A clock
   b. An accelerator
   c. A netlist

5. Simulation, formal verification, and checking tools are all
   a. Synthesis tools
   b. Design capture tools
   c. Verification tools

6. Chip testing uses input and output test
   a. Patterns
   b. Wires
   c. Faults

7. Simulation tests all the logic
   a. Gate speed
   b. Functions
   c. Tools

8. Simulation speedups include special hardware machines called
   a. Boundaries
   b. Clocks
   c. Accelerators

9. Formal equivalence checkers can test whether two _______ are the same.
   a. Netlists
   b. Delays
   c. Clocks

10. Low-power design techniques include
    a. Boundary scan
    b. Turning off sections of the chip
    c. Faster clocks

Answers: 1-a; 2-b; 3-a; 4-c; 5-c; 6-a; 7-b; 8-c; 9-a; 10-b.

7 Back-end Design Tools (Physical Design)

In this chapter...

- Introduction
- Physical Layout Tools
- Design Rule Check Tools
- Extraction and Timing Analysis Tools
- Signal Integrity Issues
- Thermal Design Tools
- Manufacturing Preparation Steps
- Product Engineering Tools
- Porting Designs to New Processes
- Summary
- Quick Quiz

INTRODUCTION

The physical design of the integrated circuit is where the design meets the silicon. Any error here will probably result in the IC not working. Examples of physical design errors include a missing wire connection or wires too close to each other.

After the FE design is synthesized, the boundary-scan registers and other test support features are inserted into the netlist. This is either the last FE step or the first BE step.

In this chapter, we discuss the physical design tools and BE checking tools. The major BE tools are highlighted in Table 7.1.

Let us join Nora again as she questions Larry, a layout designer.

Larry: Good to see you again, Nora.

Nora: You also, Larry. I hope you don't mind all my questions. I think I am beginning to understand EDA. Would you review the back-end part for me, please?

Larry: Sure, Nora. I think you know how ICs are made using a series of mask patterns. In the back-end phase, we translate the front-end transistor software models into a set of mask patterns. Like stencils, the mask patterns form areas on the silicon wafer that become actual devices and wires during IC manufacturing. Each step in the manufacturing process uses masks.

The ICs are formed in layers on the silicon surface. The upper layers are used for the metal interconnect wires. The bottom layers are used for the transistors and other devices. The mask patterns also create the wire connections between the devices, and the connections between the wires (*vias*).

(The reader may wish to review Appendix B.)

Table 7.1 Major Back-end Tools

ELECTRONIC SYSTEM-LEVEL (ESL) DESIGN	
Users	Tools
Architect	ESL Design Entry
System Engineer	ESL Modeling
IC System Engineer	ESL Verification (Test Bench)
	ESL Timing Analysis
FRONT-END (FE) DESIGN	
REGISTER LEVEL	
Users	Tools
	RTL Entry
System Engineer	Test Bench
Logic Designer	RTL Simulation
ASIC Designer	Formal Verification
Test Engineer	Design for Test
	Timing Design
	Thermal Design
	Power Design
	Signal Integrity Design
	Synthesis to Gates
GATE LEVEL	
Users	Tools
	Schematic Capture
Logic Designer	Gate-Level Simulation
ASIC Designer	
BACK-END (BE) DESIGN	
Users	**Tools**
	Floorplanning
	Layout—Place & Route
Layout Designer	**Electrical Rules Check**
Logic Designer	**Physical Rules Check**
ASIC Designer	**Extractors & Delay Calculators**
Test Engineer	**Timing Analysis**
	Power Analysis
	Thermal Analysis
	Other Analyses

PHYSICAL LAYOUT TOOLS

Larry: The actual physical creation of the chip mask patterns is called *layout.* For a house, an architect draws a detailed layout plan showing where each room, window, and pipe goes in the house. An IC layout designer "draws" where each major block, cell, gate, transistor, or wire goes on the chip.

Nora: I understand what a house floorplan is. What is the floorplan for an IC?

Floorplanning Tools

Larry: *Floorplanning* is the procedure which the IC designer uses to position each major block on the chip (like locating rooms in the house example).

While an architect is concerned about the "people flow" between the rooms, the IC layout designer is concerned about the "signal flow" between the blocks.

Initial floorplanning can be done at the FE of the design and refined at the BE. The system designer has a good idea of which major blocks need to be close together. The designer also knows which are the fastest signals and where they must travel.

There may also be external constraints on the floorplan such as chip package pins reserved for power connections. Blocks with large amounts of off-chip interconnect (like memory) often fit best in a corner. Analog blocks usually need to be isolated from the noisy digital circuits.

Automatic or semi-automatic floorplanning tools help the designer. Some tools allow initial manual arrangement (with fixed or non-fixed locations), manual intervention, and incremental changes. They allow the designer to keep the part that looks good and just rerun the rest. Other tools run only automatically, with a new trial floorplan on each run.

Here, let me draw you a sketch of floorplanning. (See Figure 7.1.)

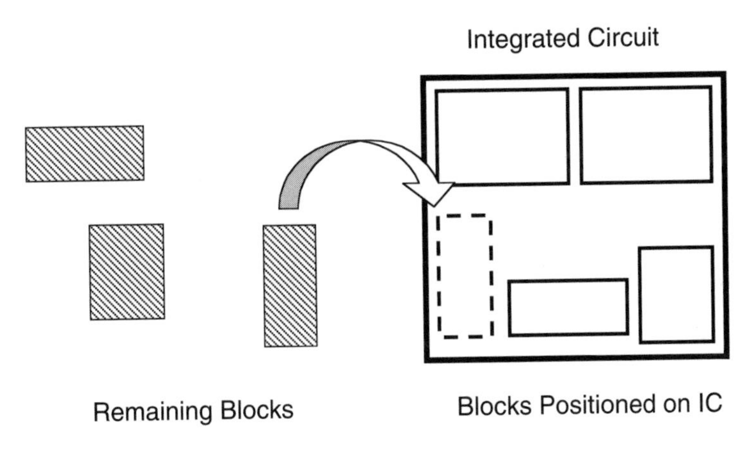

Figure 7.1
Floorplanning

Larry: Note that the floorplanning process tries out different arrangements of the major blocks. The designer tries to meet all the constraints which I mentioned and to achieve the smallest chip area.

Nora: If floorplanning locates the blocks, then what is placement?

Placement and Routing Tools

Larry: While floorplanning locates the major blocks, *placement* positions the **detailed blocks** (each cell or gate in a row or block). Again using a house example, placement is like positioning the furniture in the rooms.

Placement tools help the designer place the transistors, gates, or cells. These tools give manual assistance or run automatically. Placement tools put cells that interact a lot close together. This reduces the wire lengths when the wire routing is done later.

Nora: Okay, after you have the blocks or devices placed, you wire them together. Is that right?

Larry: Right, again. *Routing* tools search for the best paths for the interconnect wiring. This is like deciding where the plumbing and electric wiring go in a house. The routing tools try to minimize wire length, wiring congestion, and the number of wire layers. Wiring congestion is just like traffic congestion – too many things close together.

The placement of cells is very critical to how successful the router will be. Very often this is an iterative process—a trial placement is made and the routing is attempted. A designer modifies the placement if too much wiring congestion occurs or too many wire layers result. This is why place and route are usually mentioned together.

Some place and route tools allow incremental changes to be made manually. Others can only start a whole new place or route operation each time.

Although there are vast improvements in the place and route algorithms, human oversight still remains better. We humans can see the congestion problem (and solutions) globally. Most tools work (blindly) at a detailed level.

Every time a wire path changes layers to avoid an obstruction, a layer-to-layer via is created. The via adds another obstruction for the later wires to be routed. So as each new wire is added, the routing tool's job becomes more difficult.

Nora: So do designers have to go in to finish a layout manually?

Larry: Yes, quite often—the tool will do most of the work, but will not be able to complete it.

Nora: Are there different kinds of place and route tools?

Layout Styles

Larry: Yes, depending on the chip architecture. Router tools are specialized for various layout styles. There are *global* routers for wires between blocks, and *detail routers* for wires within blocks.

There are *maze* routers for general wiring, and *channel* routers for architectures with rows of cells and channels for interconnect. Complex nets and clock distribution nets use *tree* routers. Different SC architectures each need their own kind of placement tool.

Depending on the manufacturing process, there may be up to eight or more layers reserved for wiring. Some layers are reserved for power wiring, some for global wiring, and some for detail wiring.

Wires on different layers usually have different widths, spacing, and via sizes. Most digital wiring is done with the wires on each layer at right angles to the ones in the next layer. (This makes it easier for the routing tools.)

Did You Know?

The wiring is laid out in a rectangular grid (manhattan routing) named for the streets of New York City's borough of Manhattan. The manhattan wire length between two points is longer than if they are wired diagonally point to point. This affects the on-chip delays and the timing.

The X-Initiative consortium promotes *45-degree angle routing* to reduce the wire lengths and density. Many tools need to change to accommodate this. (Analog circuits are usually laid out manually and may use *all-angle wiring.*) Let me sketch you a quick example of some of the routing types. (See Figure 7.2.)

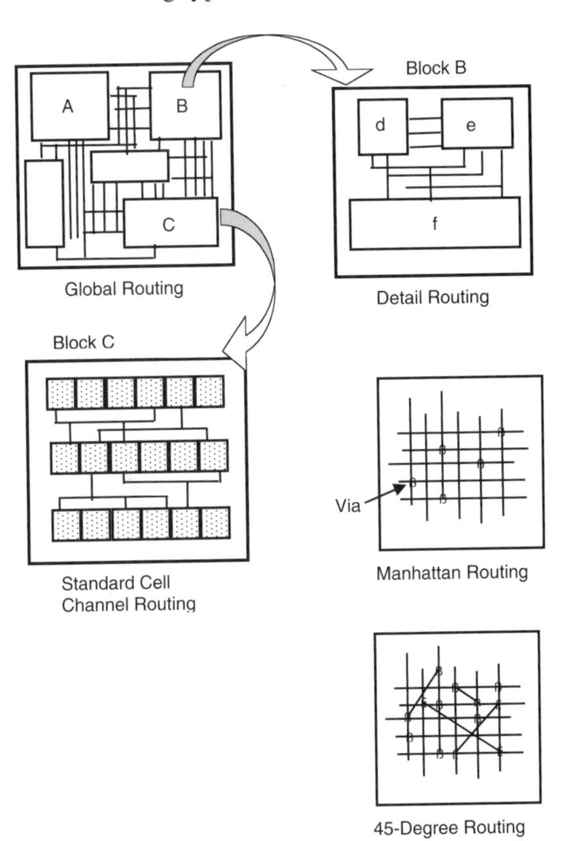

Figure 7.2
Routing Styles

Larry: Note the different styles shown in the figure. Each direction of wires is usually on a different chip layer. Global routing is between the major blocks. Within the blocks you have *detailed* routing. The blocks may be digital, analog, memory, or whatever. If the block uses standard cell architecture, *channel routing* is used to connect the cells. (Note that only the manhattan and 45-degree wiring examples show the *via* connections between the wire layers.)

Did You Know?

There is plotting software for creating large layout plots to actually see the whole layout. Different colors are used for the different layers. It looks quite impressive when you get all the wiring on a wall size sheet of paper. Many chips look like graphic art designs. (Although with increasing chip size, the plot of the whole layout is less useful for checking.)

Nora: I didn't realize that the routing could get so complicated!

Larry: Yes, there is a lot of wiring on a chip. There is about one net for every gate, and the number of gates may exceed 100M. However, the power distribution routing has even more constraints, though there is less of it.

Nora: Why is that?

Power Routing Tools

Larry: The voltage delivered on the chip must be the same all over the chip; otherwise, the chip may not work properly.

Just as in a home, large water pipes feed several smaller pipes. If the pipes are too thin, or if all the smaller pipes are using water, the pressure will drop. You need good pressure even if all the sinks and showers are in use at the same time. (So the plumbing system should be designed to ensure adequate pressure at all times.) In the water pipe analogy, water pressure is like electrical *voltage.*

Water current is like electrical *current,* and the pipe thickness is like electrical *resistance.* More resistance or more current reduces the voltage at the end of the wire. The large pipes in the house are called supply lines, and the large wires on the chip are called *power buses.*

However, large wires take up more space and reduce the area available for signal wires. Also, power buses are often restricted to just two layers. On complex chips, designers use EDA tools to rout the power buses and to calculate their required sizes (width and thickness).

Nora: So you just don't plug these chips into the wall!

Larry: Indeed not, particularly since they run at about one volt, not 110 volts! In addition, there are a lot of checks made before the design goes to manufacturing.

Nora: What kind of checks?

DESIGN RULE CHECK TOOLS

Larry: There are three main kinds:

- Electrical Design Rule checks
- Physical Design Rule checks
- Layout versus Schematic checks

There are many other checks such as timing, power distribution, thermal balance, and signal integrity. We'll get to these later.

Electrical Rule Checking (ERC) tools check that all the electrical rules are met. Examples of rules are "no unconnected wires," or that "one logic gate may drive no more than five others." ERC is also run before layout to ensure that the netlist meets all the electrical rules.

Design Rule Checking (DRC) tools perform *physical verification.*

Just as each city has different building codes and restrictions for houses, ICs have many design rules and restrictions. DRC tools check that all the hundreds of layout rules and size constraints are met. These are rules about transistor device sizing, wire width, and spaces between wires. The rules vary for each layer and are also specific to each semiconductor process.

Layout versus Schematic (LVS) checking tools. These follow the schematic or netlist and ensure that the physical layout matches the netlist one-for-one. That means no missing or extra physical nets or connections.

Those tools have to check millions of mask dimensions. One tiny error can cause an open or short circuit—and a chip failure.

Nora: No wonder you need a computer tool to check these. Someone mentioned earlier that the physical design determines the chip speed. How does that work?

EXTRACTION AND TIMING ANALYSIS TOOLS ....

Larry: Once placement and routing has been done, all the transistor and wire physical dimensions are known. *Extraction* and *analysis* tools use these dimensions to calculate key *parameters*.

Parameters include things like transistor size, wire thickness and length, and effects from nearby wires. (These were just estimated in the FE design.)

Some of these computations are very complex, involving two- and three-dimensional models.

Did You Know?

Delays are always critical to a chip's performance. The longer the delays, the slower the chip will run. Delays may also determine whether a chip operates correctly. Just getting the signals to the right place is not sufficient. They also have to get there at the right time.

Delay calculators use detailed models to calculate the critical signal delays down each path. They take into account the wires, transistor drive, and number of gates driven. There are several "standard" *delay models* in use.

The calculated parameters are entered into the design files. The dynamic and static timing checks are rerun based on physical data rather than on estimates.

Static Timing Analysis (STA) tools calculate all the gate and wire delays using agreed-upon delay models. They compute minimum, maximum, and average delays, and positive or negative *slack times.* Potential problems are listed for designer correction. STA is used in the FE design as well as after layout.

One way to speed up a critical path is to insert a transistor *buffer* circuit. This drives the signal wire faster. Inserting a buffer changes the layout, however, and some tools can do this automatically. Of course, any parameters in the affected areas need to be recalculated.

Dynamic Timing Analysis (DTA) tools simulate the chip operation with estimated delays (in the FE). They use accurate physical layout delays (in the BE). DTA tools may catch some timing problems which the STA tools miss.

Timing Closure is the procedure of finding timing problems, fixing them, and rechecking the design. Sometimes fixing one problem creates others, which then need to be fixed. With luck, the number of problems continues to decrease, eventually to zero.

However, for some methodologies there is no guarantee that closure can be found. Lack of closure can cause endless project delays and even complete failure. A picture of this is just like verification—test, find a problem, fix it, and retest. Only now the fixes involved physical changes as well.

Nora: It sounds complicated. Does it get frustrating?

Larry: Well, the timing calculations are complicated, that's true. (That's why we use EDA tools!) You're right, it can be very frustrating. And now we have even more factors to check to achieve design closure, such as signal integrity.

Nora: What is signal integrity?

SIGNAL INTEGRITY ISSUES

Signal Integrity

Larry: *Signal Integrity* measures how well a signal travels from one logic gate to another. If the signal on a wire is larger than a certain voltage threshold (say, 0.5 v.), then the gate interprets that signal as a logical "one." If the signal is below that threshold, the gate decides that the signal is a "zero." So, anything that disturbs the signal voltage level can cause an error.

Noise can reduce the signal voltage and increase the time to travel. It is similar to static on the radio and dropout on your cell phone.

Signal Integrity Analysis (SIA) tools evaluate the effect of *noise* (from all sources) on the signal. This includes coupling between signals *(crosstalk)*, and other noise on the chip. The SIA tools flag danger spots and critical nets.

All these noise sources can cause errors at the receiving gate. The error can cause a momentary (*soft*) failure in the chip operation. Signal integrity requires complex three-dimensional models to analyze the noise effects accurately.

Signal integrity tools do this analysis and avoid or fix the problem, or notify the designer.

Nora: Can any of these tools help my cell phone dropouts?

Larry: I'm afraid not. The analogy is okay, but the causes are different. The cell phones are transmitting signals over miles. The chips are transmitting signals over fractions of an inch.

Voltage Sensitivity

Larry: Electronic circuits will work over only a small range of voltage. Temperature, process variations, and noise all affect how the circuit performs. *Voltage Sensitivity* tools analyze the design to check that voltage is adequate all over the chip, under all conditions. Some chip operations use more current than others and this can affect the voltage at different spots on the chip.

Noise Margin

Larry: Earlier chips worked at five volts. We reduced chip voltages to one volt to lower power and increase speed. However, the noise problems on the chip have increased.

A logic gate expects to see two distinct signal levels (high or low) at its input. Noise can cause a one to look like a zero, or vice versa. The margin between the expected signal and the threshold voltage is called the *noise margin*.

With five-volt circuits, a gate can still recognize a signal even with a half-volt of noise on it. However, with one-volt circuits, a gate may not recognize a signal correctly with one-half volt of noise. The one-volt circuits give us less noise margin.

Nora: What can be done about that?

Buffers

Larry: When a signal travels down a wire, it loses voltage due to the resistance of the wire. (This is similar to the way in which water loses pressure as it flows down a long hose.) So long wires can lower the signal strength below the receiving gate's noise margin.

Amplifying *buffers* can be inserted into long lines to boost the signal strength. (They add more drive power, so the decrease in wire delay exceeds the added buffer delay.) EDA tools automatically calculate where buffers are needed and insert them. Some designs require thousands of inserted buffers. Buffers are used for both timing and signal integrity issues.

Switching Noise

Larry: When a signal switches rapidly, it can *induce* or *couple* a noise spike into a nearby wire. The coupling is worse the longer the wires run parallel and the closer together they are. Faster switching time also increases the noise.

If parallel wires (as in a bus) switch concurrently, the *simultaneous switching noise* effect is worse.

Clock lines also induce a lot of noise into a chip. Sudden changes in the power use can cause noise on the power bus. This noise is coupled through the power lines, the signal lines, and the chip substrate.

Nora: Does all this noise ever get off the chip to affect other chips?

Electromagnetic Interference

Larry: Yes, and *Electromagnetic Interference (EMI)* analysis tools calculate the *interference* created by the fast switching on the chip. EMI noise can affect other on-chip circuits, or radiate off-chip as RF interference.

It can also interfere with other system electronics such as low-level analog amplifier circuits and wireless telephones (e.g., when you can occasionally hear someone else's conversation on the telephone). We test for *electromagnetic compatibility (EMC)* to ensure low enough EMI level from an IC.

Nora: What else can go wrong?

Metal Migration

Larry: Most *solid-state* (semiconductor) electronics do not wear-out, like mechanical parts. However, there is at least one important similar failure mode.

High current in a thin metal wire can actually move (migrate) metal atoms, making the wire thinner. The thinning increases resistance and delay, which may cause a timing failure. Eventually this *metal migration* can cause a hard failure (break or opening) in the wire.

Power lines, heavily used signal or clock wires are the most likely places for this to happen. Thick enough wires can prevent migration from ever starting. *Metal migration analysis* tools check the design for this catastrophic problem area.

Nora: I didn't realize ICs could fail like that. Do the wires actually heat up and melt?

THERMAL DESIGN TOOLS

Larry: Not the wires, usually, but hot spots are a problem area, particularly for large chips and those used in portable products. Power is turned into heat when electrical current passes through a resistance. We can generate less power (and heat) by reducing the IC voltage and minimizing the current.

For many digital products, power usage occurs only when the transistors switch. (They use almost no power when not switching. See Appendix A.) The faster they switch, the more power they use, and the more heat they generate. High-speed chips generate a lot of heat, unevenly distributed across the chip. Transistor characteristics such as noise margin vary with temperature. Therefore, signals between different temperature logic blocks are more susceptible to noise.

Too much heat generated in a small area can cause a *hot spot*. Thermal analysis tools analyze the power and display a thermal chip map showing hot spots.

Extreme differences in temperatures on the chip may cause mechanical stress in the silicon. This may result in cracks and breaks in the metal wire interconnect.

Nora: I didn't realize the ICs could get that hot.

Larry: Yes, some microprocessor chips use 60 watts or more, and require fans to operate.

Nora: Is there anything else done before the chip is ready for manufacture?

MANUFACTURING PREPARATION STEPS ....

Merging Operations

Larry: Yes, there are a few more steps required. EDA tools help do these at the IC manufacturer or foundry. The design must be *merged* with all the underlying standard *base* mask layers. Those base layers include the *on-chip process test pattern*, the input/output (I/O) pads, and the standard *electrostatic discharge (ESD)* protection circuitry.

Nora: What does this ESD circuitry do?

Electrostatic Discharge Protection

Larry: The Electro-Static Discharge circuitry protects the chip I/O from electrostatic voltages. *Static electricity* causes the tiny sparks you create when walking across a carpet. The sparks can reach thousands of volts. Static electricity may also occur when handling the chip in the assembly and packaging phases. On-chip ESD protection circuits protect the IC from being *zapped* by spikes up to several thousand volts.

Nora: So that's what they mean when they say a chip got "zapped." Is the making of the mask part of the back-end?

Larry: It used to be, but now masks are made at a few specialized mask shops, and a set of masks can cost millions of dollars!

Mask-Making Preparations

Larry: There are several kinds of photomask machines. Some require unique data formats or handle the graphical polygons differently. The data is usually re-ordered to speed up the mask-making procedure.

Different optical techniques, such as *Optical Proximity Correction (OPC),* are also used for finer resolution. These require programs to prepare the chip design data files for the mask shop. (See Appendix B for more on semiconductor manufacturing.)

Nora: So you send off the mask data and you get chips back?

Diagnostic and Manufacturing Tests

Larry: Well, it's not quite that simple. First, we have to prove that the IC design works. The chips first need to be tested, either by the manufacturer or by a contract test provider. The design phase is not over until the first chip works. This event is known as *first silicon success*, and it is **the** crucial milestone in chip design.

Initial *diagnostic* testing on a new chip looks for design errors and their causes. The designers' test patterns (vectors) are used to verify the design behavior.

The first wafer needs to pass the *wafer test* in which all the chips are tested while still part of the wafer. If **any** individual chip works, then the basic design has been proved! Getting more chips working is (simply?) a process of *yield improvement*. Often, the design team has a party to celebrate first silicon success.

For designs done on FPGAs, however, the design can be reworked and re-implemented in the FPGA without requiring any silicon changes, since the silicon has already been proven.

We can use faster *manufacturing* tests once a chip design is proven to work. These screen out chips that fail from manufacturing issues, not from design flaws. Manufacturing tests include various electrical tests and fault checks.

In chip testing we want to ensure that every gate, wire, and via gets tested. Each test pattern should check for one or more possible failures.

Nora: How do they do that?

Larry: Most electrical faults on a chip are quite simple. As with home electrical problems, the usual issues are shorts or breaks in the connections. The electricity is either stuck on (a short circuit), or can't get through (stuck off, a break in the wire).

Chip failures (transistor, gate, wire, via, etc.) also show up as a signal not switching as expected. The signal will be either "stuck high" or "stuck low." Not surprisingly, these are called *stuck-at faults*. So if the test patterns check every chip fault (point of failure), we have a reasonably thorough test.

Nora: But how do they know they have checked every fault?

Automatic Test Pattern Generation

Larry: *Fault coverage* is a measure of the effectiveness of a set of test patterns. There are tools to measure this, and the goal, of course, is 100% (i.e., every potential fault is tested).

Automatic Test Pattern Generation (ATPG) tools create sets of test patterns to improve fault coverage. ATPG has been a major research area for years. These ATPG tests usually test for stuck-at faults and may be used alone or added to the designers' tests. Being able to compress the number of test patterns directly reduces the test time and chip cost.

However, high-speed manufacturing testers (made by different vendors) do not all use the same data formats. Therefore, the design test pattern files may need to be edited, restructured, or reformatted. The fewer the test patterns, the shorter the total test time. The goal is to thoroughly check the chip with a minimum number of tests.

The testing is thus structured (by EDA tools) for quick, gross tests first, and longer, detailed tests later. This reduces the cost of testing bad chips. Other tests are added to help quickly screen out bad parts.

IC testers check the chips while they are still part of the wafer (*wafer test*). Bad chips are marked and discarded after dicing. Good chips move on to be "assembled" in a package, and other IC testers then check the packaged IC (*assembly & test,* or *final test*).

Did You Know?

IC chips are not repairable. So any failure, no matter how small, fails the whole chip. The chip is either a "go" or we throw it out (a "no-go"). So manufacturing fault tests are called go/no-go tests. (Note: some chips, such as memories, include redundancy to be able to use the chip even with a partial failure.)

Nora: So who gets the chip to work if it fails?

PRODUCT ENGINEERING TOOLS

Larry: A good question. A *product engineer* is the intermediary between design and manufacturing. The product engineer usually guides the first silicon through manufacturing, using EDA tools.

Diagnostic tests are run if no chips work on the wafer, or not in the package. The product engineer performs these tests to find the source of chip failures. The source may be the design, the tests, or the process.

Various isolation tests and probe tools help with the diagnostic effort. Microscopic analysis with an electron beam microscope may be used. Finding and fixing errors at this phase can be a long, laborious, and costly procedure. Design, test, or process fixes are then made. All the verification and checking tools are rerun, and new mask or masks made.

These *re-spins* (re-do's) of silicon are unfortunately common. Great effort is made to achieve first silicon success, because the alternative is so expensive. Re-spin iterations may cause projects to run out of time or money before achieving success. So some projects use FPGAs for the final product, or use them to prove out a design before implementing it as a denser, faster silicon ASIC.

An *in-system* test runs the IC with the system (or EDA test bench) in which it must work. The ICs are *characterized* (how fast they run, over what voltage range, and so forth) using EDA tools.

After characterization, a long series of *burn-in* tests is done. The IC is operated at high power and temperature. This accelerates any sort of wear-out or early (*infant mortality*) failures. These tests add a measure of reliability to the production use of the design. A few EDA tools model semiconductor *aging* to help predict device reliability.

Nora: I guess waiting for first silicon must be a tense time.

Larry: Yes, it's pretty tense. A lot of these ICs have a short product life. Process improvements come about every 18 months or so. So there's a lot of pressure to move a design on to a faster process.

Nora: You mean they have to do it all over again?

PORTING DESIGNS TO NEW PROCESSES................

Larry: Pretty much. Maybe they'll add some features. Moving a design to a later process is called *porting*.

Some EDA tools help *port* (migrate) a chip design or cell library to a new process. Many processes are known as *process shrinks,* where only the feature size changes.

Suppose we go from, say, a 0.18-micron process to a 0.13-micron process. We shrink the smallest feature size (gate length). Most cell libraries are scaled to the smallest feature size. So every other transistor dimension is shrunk proportionally.

However, not all parameters (the resistances, capacitances, and other electrical properties) scale so nicely. The transistors and gates all have to be *recharacterized* to establish the new operating values.

Some EDA tools do most of the resizing and characterization automatically. They simultaneously ensure that the circuits meet all the design rules and constraints.

The automation of the resizing, recharacterizing and change control of process libraries reduces the manual work. An IC manufacturer or foundry may use several processes (e.g., 0.25, 0.18, 0.13 micron, etc.) at the same time. Some EDA companies specialize in the porting and maintenance of libraries.

Porting a design or library to an entirely new process is a more complex process. It may involve a new foundry, new process steps, or more layers—not just a shrink. The same EDA tools help with this migration and library development as well.

Analog circuits tend to use larger transistors than digital circuits. Digital circuit processes rapidly shrank below 1.0 micron (*submicron*). Processes below about 0.18 micron are referred to as *deep submicron (DSM)*.

Nora: So these ICs are always in a state of migration?

Larry: The libraries are, more so than the chips. Well, Nora, that's about all I can tell you about the back-end design tools.

SUMMARY

We discussed all the IC BE design steps and tools. We have gone from design methodologies all the way to first silicon.

Physical layout includes floorplanning, placement, and routing. Floorplanning arranges the major blocks on the chip. Placement arranges the smaller blocks or cells inside the blocks.

Routing traces out the wiring paths. There are several routing styles. Global routing connects major blocks, and detail routing handles the wiring inside blocks down to the transistor level. Channel routers are used for standard cell blocks.

BE design has many checking tools. Electrical rule check (ERC) checks electrical rules for gate interconnection. Design rule check (DRC) checks physical spacing and sizing rules for layout and placement. Logic versus schematic (LVS) checks that the physical wiring matches the schematic or HDL description.

After the place and routing, extraction tools compute the physical lengths of wires and spaces. Delay calculators use this information to predict accurate gate and wire delays. A static timing analyzer tool analyzes signal delays through gates and wires to ensure that no signal or clock is late. Clock pulses control the flow of data through the logic. Timing closure is the process of fixing all the timing violations without causing new ones.

Signal integrity tools analyze noise and interference from other signals in the chip. Low-voltage chips are more sensitive to noise. An even power voltage all over the chip is required, as the circuits operate over a small voltage range. Voltage sensitivity tools analyze the design to ensure that all points on the chip remain within the range.

Small chips with fast switching can consume a lot of power and run hot. Thermal design tools check for hot spots.

Several manufacturing steps lead to first silicon. Product engineering and diagnostic EDA tools analyze the first chips to ensure that one or more work.

Process porting tools help move an existing design to a new process.

QUICK QUIZ

1. An IC floorplan:
   a. Arranges major blocks on the chip
   b. Arranges transistors on the chip
   c. Lays out wires on the chip

2. IC routing tools:
   a. Arrange cells on the chip
   b. Rout out timing problems
   c. Plan the wire paths on the chip

3. Physical rule checker tools:
   a. Check the spacing between wires
   b. Check the number of loads a gate can drive
   c. Check the netlist against the physical wires

4. Signal delays are *not* caused by:
   a. Wire length
   b. Clocks
   c. Noise

5. Signal integrity tools measure:
   a. The honesty of a signal
   b. How well signals travel through the wires and gates
   c. How fast signals travel

6. Thermal analysis tools measure:
   a. The temperature of the wires
   b. The chip sensitivity to heat
   c. The temperature all over the chip

7. Timing closure is when:
   a. All delay faults are resolved
   b. Successful first silicon is achieved
   c. The chip shuts down due to a timing fault

8. First silicon refers to
   a. The first silicon wafer manufactured
   b. The first silicon chip which works, for a new design
   c. The first silicon chip to exceed the performance goal

9. Chips use more power when they
   a. Run at a faster clock rate
   b. Run at a lower voltage
   c. Use power buses

10. Voltage sensitivity tools check
    a. That electrostatic protection is in place
    b. That high voltage circuits are not used
    c. That the voltage all over the chip is within operating range

Answers: 1-a; 2-c; 3-a; 4-b; 5-b; 6-c; 7-a; 8-b; 9-a; 10-c.

8 Trends

In this chapter:

- EDA Design Environment Trends
- EDA Tool Trends
- Design for Manufacture (DFM) Trends
- System-on-Chip and IP Trends
- Semiconductor Trends
- Summary

EDA DESIGN ENVIRONMENT TRENDS ....

EDA is evolving rapidly to address the many new design issues facing the chip designer. We mentioned a few of these issues previously. The trends associated with the overall design environment include:

- Integrated Design Suites
- Run-Time Control
- Distributed Design
- System Design Links to Chip Design

Integrated Design Suites

Design tools are moving from (possibly incompatible) point tools to integrated suites of compatible tools. Some integrated suites simply use a compatible *data transfer format* to move data from tool to tool. Others use a common *data structure* or *database* to allow tools to share access to common data.

Interface formats for files or data may be *proprietary* or *non-proprietary*, *open* or *closed.* Proprietary means a vendor owns and controls it. Non-proprietary means the format is published and available to the public. *Open* means that it is free or for sale to the public. *Closed* means that it is available only with tools of the vendor.

Run-Time Control Tools

A design flow involves many tools, many steps, and many files. Every change to a file requires a revision number change to the file name. After a design error is detected and fixed, many related files may need updating as well. (Very often, fixing one error affects several files. Any steps we previously ran may also need to be rerun after a design file is changed.)

With all the interruptions to fix errors, it is easy for the designer to lose track of where he is in the design flow. Where are we in the flow? What is the state of the design? Which files were not affected, which files need to get updated, and which design flow steps need to be re-run? These issues are called *run-time control.*

Increasingly, design groups are using EDA *run-time control* tools. These tools include features that guide an engineer through the design flow. They include automatic revision control and tracking of files (possibly) invalidated by a design change. They reduce errors from the inadvertent mistyping of a file name.

Some tools handle automatic launching of a tool and its associated data, parameters, and output files. The tools provide automatic iteration tracking, display a history log of all the design steps and files involved, and so forth.

Distributed Design

EDA tools are evolving to support design teams separated by time, geography, language, and culture. Some issues include sharing enormous design data files, revision control, and backup.

Backup is particularly important since a damaged or lost file is very costly. Re-creating a file requires a great deal of work. The designer may have to repeat many steps in the design flow to re-create the lost file.

Other distributed design issues are security and fast file transfers (both incremental changes and whole files). Video-conferences, net meetings, and long teleconferences supplement the non-EDA side of the challenge.

System Design Links to Chip Design

Traditionally, system engineers do electronic system level design and chip designers do IC design. They exchange information manually, since their tools do not communicate.

However, several groups are trying to connect *system level* design directly to chip level design. Some emerging system level design languages link directly into the chip design system. These include extensions of the C and C++ programming languages, or the Verilog HDL. They are competing to be recognized as a standard.

Other groups are trying to develop a high-level, *executable specification* language for either hardware or software. One could create hardware description code or software code directly from the design specification. Related work includes ways to describe system design *constraints* (such as size, power, or speed).

Did You Know?

Some system designers prefer using a programming language to describe the system. Others prefer a hardware description. There are heated debates over which is the best approach.

Their preference tends to follow their experience: C programming experience, or graphical and HDL experience.

Hardware designers are used to thinking about concurrent systems, where operations are going on in parallel. In contrast, most programming languages are inherently sequential. Therefore, most programmers have learned to use a sequential design approach (even if a language has recent extensions to support concurrency).

EDA TOOL TRENDS

Tool trends include:

- Design Closure
- Formal Verification
- Design Repair
- Design for Test

Design Closure

The point when all open (unsolved) design problems are closed (fixed) is called design closure. (This is similar to unsolved crimes—the "case" is open until the criminal is caught. Then the case is closed.)

With DSM chips, the various design problems interact. Fixing one problem may cause a new one somewhere else. For example, to solve a thermal problem the designer may have to lengthen a signal path. The longer wire may increase a critical delay, causing a timing failure. Design closure may thus take a long time, with no guarantee it can even be reached.

New tool suites are attempting to balance and optimize interacting problems concurrently instead of sequentially. Some individual tools now estimate physical design issues earlier in the design flow, to avoid or sidestep the problems before they arise.

Formal Verification

As design complexity increases, verification becomes more time-consuming and expensive. Some design entry languages support notes from the designer about the *design intent*, i.e., what the design should do and what it should not do. These expectations (*properties*) are inserted into the HDL code (*assertions*) or are provided as a separate file.

Then a *formal verification* program can check the design for consistency with the design intent notes. The formal check is quick and thorough and can catch some errors that simulation does not.

Design Repair

Some BE tools can correct design rule failures automatically as well identify them. Previous tools could flag the problem, but the designer had to fix it manually. Automatic repair really helps when a tool finds some 10,000 instances of a physical design rule error (e.g., a physical spacing error on a transistor which is used thousands of times in the chip).

Design for Test

When chips were simpler, designers often left the testing up to the manufacturing test engineers. There is more interest in Design for Test (DFT) as the cost of testing increases. DFT requires the designer to plan for all the testing required from idea through manufacture.

The designer must design-in support for all the testing, right from the beginning. Several EDA tools are available to assist with DFT approaches.

***Built-in Self Test* (BIST).** This uses a set of test patterns to test the chip at speed. The patterns are generated on the chip itself. As test time and equipment become more expensive, BIST has become more cost-effective. *Logic BIST* tests the logic blocks. *Memory BIST* is used for the self-test of memory blocks.

Automatic Test Bench Generation. This helps the engineer easily create tests. The final verification of the design is only as good as the test cases developed. Manual test generation is tedious and error-prone. Test bench tools provide a whole environment to help create, order, and run tests.

Testability. More transistors and functions (complexity) require more test patterns to test. The time and cost to adequately test a chip are increasing dramatically. Analog and RF blocks require special test equipment, usually run manually.

Several tools can evaluate a design to grade its testability. Some can insert test points and flip-flops to improve the chip testability and reduce the number of tests required.

Reliability. The probability of chip errors increases dramatically as the number of blocks grows. Suppose we are 95% confident that each block works. What can we expect when we put 15 blocks together on a chip? A 50% likelihood that it will fail! So the total testing cost for each block on larger chips can increase very steeply.

Frequency. A chip may be running a thousand million operations a second. If the error rate is only one in a hundred million, there could still be 10 errors a second! The problem of chip failure thus grows not only with complexity but with increased speed.

We may tolerate a failure in our personal computer or cell phone every few hours, but not in our car engine or airplane!

DESIGN FOR MANUFACTURE (DFM) TRENDS..........

With each reduction in IC feature size, additional manufacturing issues arise. These include increased complexity, higher clock frequencies, power dissipation, component density, and reliability.

The effects of the manufacturing process on the design need to be understood. Chip designers, lithography, mask-making, processing, and test engineers need better two-way communication.

More manufacturing-aware tools are coming onto the market. These tools take into account manufacturing tolerances, voltage sensitivity, and technology constraints. Considering manufacturing early in the design process can significantly improve the eventual semiconductor chip yield.

Significant DFM trends include:

- Design Redundancy
- Chip-to-Chip Differences
- Mask Enhancements

Design Redundancy

One trend is greater use of *redundancy* to achieve a higher manufacturing yield. This approach has been most successful with memories and FPGAs. Memory blocks incorporate spare memory cells used for self-repair of hard failures. Redundancy also enables repair of *soft errors* (errors due to noise or manufacturing tolerance variations). Now similar techniques are being applied to the logic circuits.

Chip-to-Chip Differences

Another trend is to include the manufacturing process variations in design analysis. Chips can differ from one another, even when made on the same wafer. (As in baking a batch of cookies, each cookie is unique.)

The analysis uses *defect density* information (the predicted number of defects at each manufacture step). It identifies probable failure due to the normal manufacturing variations. The analysis applies within the chip, and from chip to chip across the wafer.

One example is clock distribution which must span a large chip area. Manufacturing variations can cause clocks to arrive at the wrong times. The designer uses circuit design and layout analysis to reduce the effect of expected manufacturing variations.

Mask Enhancements

Will this incredible shrinking ever end? The end has been predicted repeatedly, but so far it hasn't happened. The original National Technology Roadmap for Semiconductors predicted 70 nm (0.07 micron) feature size by 2010. A 50-nm process was announced as early as 2002. (However, people are encountering many new problems at 90 nm and below. New transistor models and tools are required to account for these new issues.)

Whereas mask costs used to be relatively low (about $3,000), mask sets (all the layers) now run over $1,000,000. A SEMATECH survey showed that as few as only **six** wafers were run for some ASIC parts. At that rate, the chips must be priced at several hundred dollars just to break even. Some industry observers predict the end of ASICs due to this low volume and high mask cost. (Of course, the FPGA marketing folks have been predicting the end of ASICs for years...)

There is still life in the optical lithography used to make the manufacturing masks. Various optical tricks (*Resolution Enhancement Techniques*) are used to improve the resolution and mask image. Two main approaches are *Phase Shifting Mask (PSM)* and *Optical Proximity Correction (OPC).*

These techniques require EDA programs to prepare the chip design data files for the mask shop. Currently, the tightest specifications apply to everything on the mask, whether critical or not. There is no range or scale for defects.

The IC designers must learn and adjust for the mask maker's constraints. Alternatively, the IC designer's intent must be passed down to maskmaking and manufacturing. This would allow variable specifications on the masks to be implemented. It is not clear which approach will be taken.

SYSTEM-ON-CHIP AND IP TRENDS ....

A major CEO recently noted that his company was designing three times more complex chips in only two-thirds of the time, compared to only a few years previously. Trends in *system-on-chip (SoC)* and IP development which affect EDA tools include:

System-on-Chips (SoCs). Larger complex chips now include most of the system electronics. This is an increasing trend. System chips usually include multiple memory blocks, one or more microprocessors, logic, and several I/O interfaces on the same chip. (See Appendix F for more on SoCs and IP.)

Design Re-use. The requirements of short TTM have driven the increased re-use of existing design blocks. Integrating these blocks into SoCs is changing the design methodology. (See Appendix F.)

Intellectual Property (IP). The design, acquisition, and use of IP blocks have become an emerging industry. IP includes memory blocks, microprocessor blocks, I/O interfaces, RF, analog, and programmable blocks. Concatenation of smaller IP blocks into larger function IP blocks is also increasing. The integration of blocks from different EDA design methodologies is a major issue.

Memory Blocks. The types and sizes of customizable memory are increasing. Many system chips have multiple blocks of memory. Some chips are 70-90% memory. So more EDA tools are being developed to model multi-memory systems and interfaces.

On-chip Buses. There is increased use of on-chip buses to simplify interfaces between IP blocks.

Customized Processors. These are becoming more prevalent. Several companies provide IP architectures to support custom-made processors. These sometimes have special instructions and architecture for a specific application.

SEMICONDUCTOR TRENDS ....

As we have seen, the semiconductor industry includes several different players and types of ICs. The continuing smaller feature sizes have created larger problems. EDA tools help solve the IC problems. So IC developments have had great influence on the development of EDA tools. It is reasonable to ask, "What next?"

Many new design issues arise as the process dimensions shrink below 0.18 micron, and again at 0.90 micron. Performance, power, thermal design, and physical design are affected. New materials are emerging. These developments affect the transistor models and manufacturing rules used in the EDA tools.

The trends described below include:

- Performance Design Issues
- Power and Thermal Design Issues
- Physical Design Issues
- New Materials and Lithography

Performance Design Issues

Interconnect Delay. Thinner wires have more resistance and capacitance, which increases the delay through them. (See Appendix A.) A larger chip means that some wires will be longer—further increasing the delay. This increase in delay conflicts with the attempt to run the chip at faster speeds.

Lower Voltage and Smaller Transistors. A smaller transistor and lower voltage give a shorter delay (good—it switches faster). But it also makes a weaker signal (bad—more signal wire delay and sensitivity to noise).

Crosstalk. Rapidly changing signals on one wire can induce a transient signal on nearby wires. If many signals are changing nearby at the same time, the effect can be large. This simultaneous switching noise can cause signal delays and logic or memory errors.

Signal Integrity. Noise affects the integrity of the signal (how well it is defined as a "1" or a "0"). Noise comes from many sources (clocks, crosstalk, power, and ground lines, or chip leakage paths). Noise may cause signal delays, or logic and memory errors.

Clocks. With more functions on the chip, there may be hundreds of clocks running all over the chip. The noise generated by the clocks also affects the signal integrity.

RF Blocks. Radio frequency (RF) chips used to be separate ICs. Wireless applications are an increasing trend. Wireless circuitry embedded on the digital chip is also an increasing trend. RF circuits run at very high frequencies. The RF signals do not stay on the wires but can radiate throughout the chip. They may induce noise and cause memory upsets affecting signal integrity.

Power and Thermal Design Issues

Power Distribution. More transistors switching need more power. Power distribution to different logic blocks on the chip grows more complex. Thinner wires have more resistance, increasing the voltage drop to the blocks.

Electro-migration. With smaller wires, heavy currents can actually move metal atoms. This thins the wire even more, increasing the current density. Eventually a wire break results.

Thermal Distribution. With smaller feature sizes, power may be focused in dense hot spots on the chip. This affects the local transistors' speed and noise sensitivity. It can also cause mechanical stress and cracks in the wires or the silicon.

Physical Design Issues

Antenna Effects. Certain layout patterns can cause long unwanted wire stubs which affect the signal integrity.

More Wiring Layers. With more transistors, more layers of interconnect are needed to wire them together. Each layer of interconnect adds more vias and more potential manufacturing problems (as well as cost).

More Wiring Vias. Vias (pass-throughs from one wiring layer to another) are weak links in the manufacturing process. More vias mean more chances for a connection failure.

Mask Costs. With more layers, there are more mask costs. (Each DSM mask set now costs millions of dollars.) An error which requires a new mask set is very expensive in time and dollars!

Hot Electron Effect. The small dimensions can create high electric fields which accelerate electrons. These hot electrons can damage the transistor. This is a major reliability problem.

New Materials and Lithography

New materials. The use of silicon is being challenged as the all-purpose substrate for integrated circuits. Silicon is cheap and easy to process. Silicon Germanium (SiGe), Gallium Arsenide (GaAs), and Strained Silicon are better for high-speed circuits.

Silicon-on-Insulator (SOI). This is one way to reduce transistor delays and interactions through the chip substrate. (A layer of silicon dioxide insulates all the active transistors from the base chip.) SOI reduces leakage, power loss, and enables faster circuits.

Copper. This has replaced aluminum for faster interconnect wiring. New (low K) insulating materials have also been used to reduce interlayer capacitance for faster interconnect. These trends affect the EDA interconnect and switch delay models.

Finer Lithography. The optical lithography may come to an end, but scientists are looking **beyond that.** The trend toward further reduction of feature size may come from the use of X-rays or electron beams. New work is assembling patterns by directly moving molecules and even individual atoms. New EDA tools will no doubt be needed to deal with these developments.

SUMMARY

We have examined EDA tool and semiconductor trends.

We looked at the several developments in the design team environment and in individual tools and suites. These included integrated design suites, run time control, distributed design, and system-chip design languages.

EDA tool trends include design closure, formal verification, design repair, design for test, and memory system design tools. The cost of test is increasing rapidly as the chip complexity grows.

EDA tools are including manufacturing aspects early in the IC design. Design for manufacturing trends include design redundancy, chip-to-chip differences, and mask enhancement tools. Mask costs are increasing with no end in sight.

The semiconductor trends have and will have great influence on the EDA tool landscape. Whole systems on a single chip have become common. These SoC chips have several EDA-related issues, including design re-use, IP, and integration of blocks.

Other chip issues include performance design, power and thermal design, DSM physical effects, and new materials under development. All of these involve EDA tools for analysis or checking.

Appendix A

Elementary Electricity

In this appendix ...

INTRODUCTION

Electronic Design Automation (EDA) provides software tools to integrated circuit designers. The reader needs some familiarity with ICs to grasp what the EDA tools do.

In addition, understanding ICs requires an introduction to basic electricity and semiconductors. This chapter introduces electricity.

Atoms and Electrons

All matter is made up of tiny particles called atoms. Atoms are the building blocks of every kind of material—water, steel, and skin. We think atoms consist of a center *nucleus*, and a cloud of *electrons* swarming around the nucleus. (The model looks like the sun surrounded by many planets.)

Figure A.1 shows electrons swarming around a nucleus. Note that the electrons are much smaller than the nucleus. (Also, the picture is not to scale.)

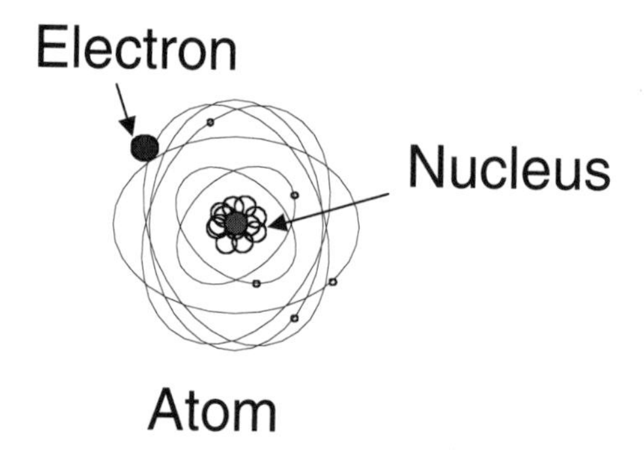

Figure A.1
Atom and Electrons

While the planets are attracted to the sun by gravity, electrons are attracted to the nucleus by an *electric charge*. Electric charges occur as *positive* and *negative*. Like charges repel and opposites attract. (Similar to male and female, but without sex!)

Electrons are negatively charged and the nucleus is positively charged, so they attract each other. The force between charged particles is stronger the closer they are to each other. The more electrons there are together, the stronger the force.

Conductors, Insulators, and Semiconductors

In some materials, most electrons are tightly bound to the nucleus and cannot move around much. These materials (e.g., glass, wood, plastic) are called *insulators*.

In other materials, electrons are loosely bound to the nucleus and move readily. These materials, mostly metals (e.g., gold, copper, or aluminum), are *conductors*. Electrons move (current flow) easily from atom to atom.

Between insulators and conductors, there is a middle class of materials called *semiconductors*. Semiconductors such as silicon, germanium, and gallium arsenide are uncommon materials to most people. They conduct electron current poorly unless we can inject them with extra electrons. This is exactly what we do!

ELECTRICAL ATTRIBUTES ....

Electrical Current

We can force electrons to move in a conductor (along a copper wire, for instance). We do this by pushing more electrons in at one end and removing them at the other. A *source* (battery or generator) pushes electrons through a wire and back to the source.

When an electron leaves an atom, it leaves a positively charged *hole*. We can think of electrical current as holes moving in one direction or electrons moving in the opposite direction. By convention (thanks to Benjamin Franklin), positive current flows from plus to minus.

Figure A.2 shows a *current source* (like a battery) with a wire connected to a *current sink*. The current returns to the source by a return path (wire or ground). Conventional current flow (the big arrow) flows from plus to minus. The electrons actually move from minus to plus.

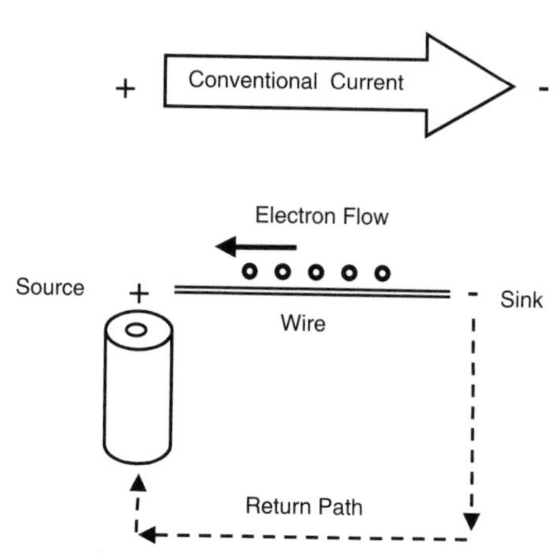

Figure A.2
Electrical Current

Electricity consists of a current of electrons flowing in a circular path. It may pass through a network (circuit) of wires and electrical components. (The word *circuit* derives from this circular flow.) More electrical current means that more electrons are flowing.

Electrical current flow is measured in *amperes* (*amp* is short for ampere). A kitchen toaster may use about one ampere to burn your toast. A flashlight uses a few *milliamperes* (1/1000 of an amp, *ma* for short) to light a lamp. Circuits on an integrated circuit may use only a few *microamps* (millionths of an ampere).

Electrical Voltage

Higher water pressure in a hose forces more water through the hose. Similarly, the higher electrical pressure (measured in *volts*) forces more electrical current through a wire.

Normal house *voltage* (in the U.S.) is about 110 volts. Flashlight batteries may supply one to five volts, depending on the size.

ICs are powered by one to five volts. A small electrical voltage or current use to convey information is called a *signal.* Wireless radios and cellular telephones are concerned with *microvolt* (one millionth of a volt) signals.

The positive terminal on an electrical source is called *hot, power, positive,* or *Vdd.* The negative terminal on an electrical source is called *ground, return,* or *negative.*

Resistance

A long, thin hose has more resistance to water flow than a short, thick hose. The pressure will drop more in driving the water through the high resistance hose.

Similarly, electrons encounter more *resistance* through a long, thin wire than a short, thick wire. Some materials have less resistance to electron flow than others. For instance, gold has less resistance than aluminum.

Capacitance

Imagine a row of buckets connected by sections of water hoses. The water current must fill each bucket up before it moves on to the next one. It takes longer for water to get to the end than if there were no buckets.

Similarly, a circuit can store electrons in many places distributed along the wire path. The electron "buckets" are called *capacitance,* measured in *micro-* or *nano-farads* (millionths or billionths). Any two conductors separated by an insulator layer form a storage *capacitor* for electrons.

The combination of resistance and capacitance can slow a bunch of electrons (signal) flowing down a wire.

Inductance

Inductance in a wire acts like inertia. It resists the attempt to slow or speed the current flow in the wire.

Direct and Alternating Current

Electricity from a battery source flows only in one direction, called *direct current (DC)*. Electromechanical generators use rotating magnetic fields to move electrons. Their current may be DC, or it may surge alternately back and forth, called *alternating current (AC)*.

Most household current is AC since it is easier to distribute than DC. *Transformers* convert AC voltage from thousands of volts (power lines) down to 110 volts for home use. Household electricity alternates back and forth 60 times per second (60 Hertz, or 60 Hz for short).

OTHER ELECTRICAL EFFECTS

Static Electricity

Electrons can get separated from atoms and accumulate on insulators. The resultant charge can build up to thousands of volts. Static electricity causes the little sparks one produces from walking across a dry carpet. The repelling force of the electrons causes the "hair standing on end" effect. Static electricity also causes lightning, but with much higher voltages and currents (millions of volts and amperes).

Coupling

An electrical current in a wire close to another wire can affect the electrons in the second wire. This coupling is due to the electrons' repelling force and to the electromagnetic effect of moving electrons. The faster current changes occur in the first wire, the more effect there is on the second wire. This effect is called *coupling* or *crosstalk*. The closer together the wires, the greater the coupling.

Waves

AC current can made to alternate at higher speeds (Hz). At higher speeds the current begins to radiate radio waves (*electromagnetic waves*). These waves can cross empty space and do not need to move in wires.

Varying the AC voltage changes the height (*amplitude*) of the wave. Information can also be transmitted by varying (*modulating*) the *frequency* of the waves (*Frequency Modulation—FM*).

Amplitude Modulated (AM) radio waves are used at a few thousand hertz (*kHz*). *Frequency Modulated (FM)* waves are used at higher frequencies. Broadcast radio uses both AM and FM frequency bands. Portable and cellular telephones use still higher frequencies (e.g., 900 million hertz (*megaHz*) to five billion hertz (*gigahertz*)).

Radar, infrared (radiant heat), visible light, ultraviolet light, and X-rays are all higher frequency electromagnetic waves.

ELECTRICAL COMPONENTS

Besides wires, individual components exhibit resistance, capacitance, and inductance. These devices are called resistors, capacitors, and inductors.

Other semiconductor devices called *transistors* switch the current on and off. They can also *amplify* electrical signals (use a small voltage or current to control or modify a larger one). A circuit wires these components together to perform a useful function (e.g., a switch or amplifier).

Semiconductor Devices

Most semiconductor devices are made from silicon. Pure silicon does not conduct electrons very well. However, if a few foreign atoms are *doped (implanted* or *diffused*) into the silicon surface, that region will conduct better.

Boron atoms contribute more holes to form *P-type silicon*, which conducts. *Phosphorus* atoms contribute more electrons to *form N-type silicon,* which also conducts.

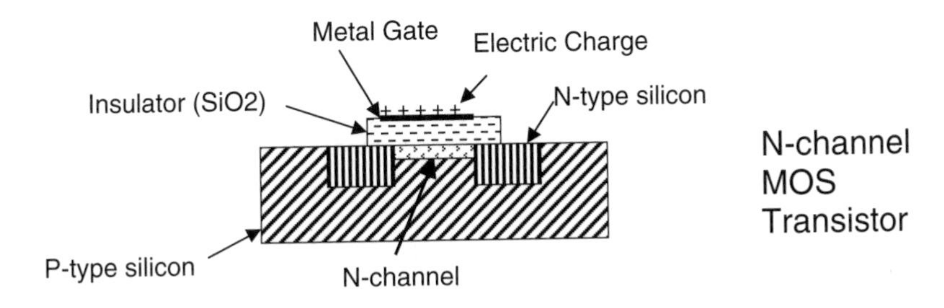

Figure A.3
MOS Transistor

One kind semiconductor structure is a *metal-oxide-semiconductor (MOS)* transistor. Figure A.3 shows a simplified picture of this structure. It has heavily doped N-type silicon areas called a *source* and a *drain*. A P-type silicon gate separates them. Over the gate is a thin layer of insulator (SiO2 or silicon dioxide), with a metal layer on top of that.

If the metal layer receives a positive charge, it attracts electrons to a thin *channel* just under the insulator. This channel forms a conductive path from the source to drain. The device is called an *N-channel MOS transistor.*

A similar structure is called a *P-channel MOS transistor.* A signal to the gate of both kinds of transistors turns one on and the other off. This complementary action is reflected in the name of the circuit—*Complementary MOS (CMOS)*.

CMOS circuits are widely used because they are simple and consume little power when on or off. (They consume power only when switching between on and off.)

Various other materials and structures are used to make faster, lower-power, or higher-voltage transistors.

Appendix B

Semiconductor Manufacturing

In this appendix...

INTRODUCTION

We discuss IC manufacturing briefly because many of the IC design issues and terms stem from the manufacturing process. The EDA tools that support the design deal with the same terms and issues.

Integrated Circuits (ICs) are made from semiconductors. *Semiconductors* are materials which conduct electricity poorly. They conduct better than insulators (such as glass), but less than conductors (such as gold or aluminum).

Most ICs are made from *silicon*, a brittle metallic-looking material. Silicon is the main ingredient of common sand (quartz, made of silicon and oxygen).

The basic IC building block is a *transistor.* Semiconductor manufacturing technology can create of hundreds of millions of transistors per chip. The process *integrates* both transistors and *interconnect* (wires) on the same IC.

ICs are made in wafer fabrication plants (*wafer fab* or *fab* for short) by semiconductor companies or *foundries.* Foundries do only wafer fabrication, while semiconductor companies do design, test, and fabrication.

MANUFACTURING PROCESS

Pure silicon is grown from single crystals into long cylinders 6 -12 inches in diameter. Diamond saws then slice the cylinders into very thin *wafers.*

A process of masking, etching, and diffusing steps creates hundreds of identical ICs on each wafer. Precise ovens process batches of wafers at over 1,000 degrees centigrade.

Figure B.1 shows a simplified overview of the basic manufacturing process steps, which are listed below.

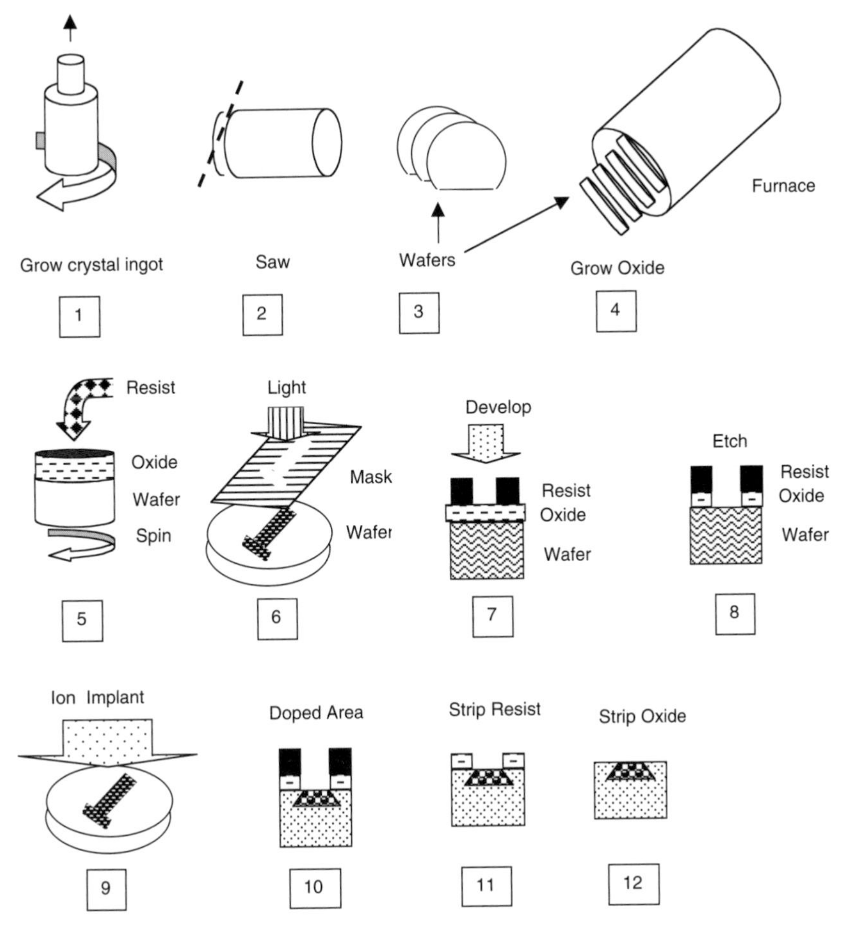

Figure B.1
Integrated Circuit Fabrication

1. Grow a pure crystalline silicon ingot
2. Saw the ingot into wafers with a diamond saw
3. Polish wafer and flatten one edge of the wafer for alignment
4. Grow a protective layer of silicon dioxide (quartz glass) on the wafers
5. Spin on a very thin film of light sensitive photoresist (resist)
6. Expose the resist with ultraviolet light through a mask
7. Develop and rinse the resist, leaving the pattern on the wafer
8. Etch the pattern into the oxide layer with acid, stopping at the silicon

9. Implant atomic impurities (*doping*) to make P-type or N-type silicon (ion implantation or high-temperature gas diffusion)
10. Note that wafer now has patterns of doped semiconductor areas
11. Strip off the resist
12. Etch off the oxide

Repeat steps similar to 4-12 for each layer (12-20 layers for a CMOS process).

Note that the masks pattern small areas in layers on the silicon surface to make transistors. Aluminum or copper metal layers are formed and patterned similarly to make the interconnect wires.

EDA tools create all the mask patterns for each layer during physical design. Most of the wafer thickness is for mechanical support. All the circuitry is formed at the surface of the wafer.

Masks and Feature Size

Photolithography is the mask-making technology used to pattern the ICs. Light is focused through fine mask patterns onto the wafer surface.

One of the transistor parts (the transistor *gate length*) is the smallest pattern used. All the other chip dimensions are scaled up proportionally to this minimum size *feature*.

We refer to a particular process by the size of the smallest feature we can make using it. Thus a 0.13-micron semiconductor process refers to a minimum feature size of 0.13 micron.

Process technology has reduced the feature size from about 10 microns (millionths of a meter) to less than a micron (*submicron*). We refer to sizes below 0.18 micron as *deep submicron* (*DSM*). At this level, many additional problems arise for EDA tools to address.

How big is a micron? A micron is one millionth of a meter. Spider silk is about as fine a thread as the eye can see unaided. It is about two or three microns in diameter, about 1/30th the size of a human hair. (Appendix D has more size comparisons.)

Manufacturing Test

Large machines test the chips on the wafer with arrays of needle-like probes. EDA tools help develop the manufacturing tests used. Chips that fail are marked with a dot of ink.

The processed wafer is then scribed and broken into individual IC chips or *dice* (about 1/16 to 3/4 inch on a side). After dicing, the bad dice are discarded.

Packaging

Semiconductor chips are quite fragile and moisture or oxygen can easily contaminate, damage, or corrode their surface. Hence, we seal them in ceramic or plastic packages.

Fine gold or aluminum wires connect the tiny chip *pads* to the larger package *pins*. The wires are only about one mil in diameter. (A mil is 1/1000th inch, less than the thickness of a human hair).

There are many package types, pin arrangements, and acronyms associated with packages. Some EDA tools deal with the power dissipation, temperature, and timing associated with the packages.

IC TESTING

EDA tools check timing *on* the IC and *between* ICs on the PC board. The package contributes to the interchip delays and affects the power and heat distribution.

After assembly, the packaged chip is given a final test through the package pins. It is important to test the chip at full speed. This means that the test equipment must be able to run faster than the chip it is testing. This can be very difficult and expensive for leading-edge chips.

IC test machines cost millions of dollars. The time involved in testing is a very significant (and increasing) part of the final chip cost. EDA tools are used to make the tests as short and fast as possible.

Note that one defect among millions of transistors and interconnections can cause the chip to fail. The percentage of good chips on the wafer is known as *yield,* and the chip cost is directly dependent on it. Yield is a critical metric in semiconductor manufacture.

PROCESS IMPROVEMENTS

The diameter of IC wafers has increased from two inches to about 12 inches. This enables many more chips per wafer, with lower cost per chip.

The number of transistors per chip has doubled about every 18 months for decades. (Intel's Gordon Moore made this observation, referred to as *Moore's Law,* in 1965.)

Process technology improvement has reduced the cost per transistor and increased the complexity of the IC. Every year some industry pundit predicts that Moore's Law will end within five years. Both the technology improvements and the predictions of its imminent end have persisted for decades.

Appendix C

Signals to Software

In this appendix...

- Introduction
- Transistor Circuits
- Analog and Digital
- Memory
- Logic
- Signal Delay
- Computers
- Software

INTRODUCTION

In Appendices A and B, we introduced basic electricity and semiconductors. In this appendix, we cover additional concepts needed to understand integrated circuits, EDA tools, and other terms.

TRANSISTOR CIRCUITS

A *signal* is the presence of an electrical voltage or current. We mentioned how transistors work to amplify small electrical signals or switch them on and off. With transistors, resistors, and capacitors, engineers wire together circuits to amplify signals and do logical operations. Let us review several functions of these transistor circuits.

- Switches—turn on and off
- Buffers/Drivers—drive signals faster on wires or cables
- Clock circuits—generate and drive clock pulses to synchronize computer operations
- Gates—logical operations (AND, OR, compare, add, subtract, etc.)
- Memory and flip-flop circuits—remember a “1” or a “0” value

(Most ICs today use a variety of complex gates chosen from a library of gate and flip-flop circuits. However, they all perform some variation of these simple functions.)

- Radio Frequency Amplifiers—amplify high-frequency, low-level signals
- Power converters—convert high voltage to low voltage, or vice versa
- Analog-to-Digital Converters (ADC)—change analog values into digital numbers
- Digital-to-Analog Converters (DAC)—change digital numbers into analog signals

ANALOG AND DIGITAL

Analog

The real world is largely continuous shades of gray, not just black and white (digital). Anything we measure in the real world is usually analog. Temperature, voltage, and water pressure are all examples of things which vary in a smooth, continuous way.

Most electrical signals from temperature, thermal, and other sensors are *analog signals*. These vary smoothly from large to small, low to high, slow to fast.

Analog circuits work with these continuous signals. They deal with amplifying very small signals, screening out noise, controlling power, and so forth.

Digital

Meanwhile, another view of the world exists in discrete chunks. Money, shoes, and pages in a book are examples of discrete things. Electrical signals can be restricted to two, four, or eight voltage levels, for example. Circuits that work with discrete levels are called *digital circuits.* Because we used to count with our fingers (*digits*), anything we count is usually *digital.*

When we count, we use only ten symbols (0–9) to represent all possible numbers. We do this with a row of symbols. Each **position** in the row has a different value or weight. The first position has a weight of one; the second position to the left has a weight of ten, and so on. This is the *decimal system* of number representation, based on ten. Computers use a *binary* number representation system, based on two.

Two is a useful base because there are only two symbols to deal with: 0 and 1. It is easy to make circuits that work with only two levels of voltage. These levels (high or low, on or off, 0 volts or 5 volts) represent the two symbols, 0 and 1.

Two level digital circuits are easier to build than precision analog circuits. They are also less sensitive than analog circuits to noise, voltage variations, etc. That is why so much of electronics is digital.

Analog and Digital

Most systems have connections to the real world of analog signals. Many ICs have analog circuit blocks to interface with real-world signals from sensors or wireless sources. Some *mixed signal* blocks have both analog and digital circuits. These typically are *analog-to-digital converters* (ADCs), or *digital-to-analog converters* (DACs).

An *analog-to-digital converter (ADC)* can convert an analog voltage into a digital number. An ADC can *sample* a signal voltage at frequent intervals and store the number for each sample in memory.

Voice and music signals are transformed into analog signals by microphones. An ADC converts the analog signal from the microphone to digital numbers. A video chip converts images into digital dots (pixels).

Conversely, a *digital-to-analog converter (DAC)* can convert a digital number into an analog voltage. A DAC can convert digital numbers from, say, a *compact disk (CD)* into an analog signal. The analog signal can then drive a loudspeaker, reproducing the music.

MEMORY

A valve or switch stays on or off when you turn it or switch it. It "remembers" its last position. Similarly, we can make *memory* circuits which remember their last setting (on or off, high or low).

Each individual memory circuit is called a *memory cell.* We can use a memory cell for each position in a number. Then a row of cells can "remember" a number.

One kind of memory cell that can be set high or low is called a *flip-flop*. A flip-flop can thus remember a "1" or a "0" (i.e., one *bit* of information). The information in the cells is called the *state* of the machine. Whenever one or more bits changes, the *machine state* changes.

A row of flip-flops is called a *register* and can hold eight bits (a *byte*) or more. A word may be anywhere from eight to 64 bits (or more) long, depending on the design of the machine.

Arrays of circuits that hold thousands and millions of bytes are called *memory blocks* or *memory banks.* The bits of information stored in memories are called *data.* Data can be any kind of information such as numbers, text, speech, music, or video.

LOGIC

Logic is the circuitry used to compare, add, or subtract data words. A *logical gate* circuit typically has many inputs and just one output. Gate circuits are built of simple electrical switches. Logic is thus decision-making with switches.

See Figure C.1. A table next to each gate symbol shows the gate output for each set of inputs. The logic gate example shows gates with only two inputs, but gates may have three or more inputs.

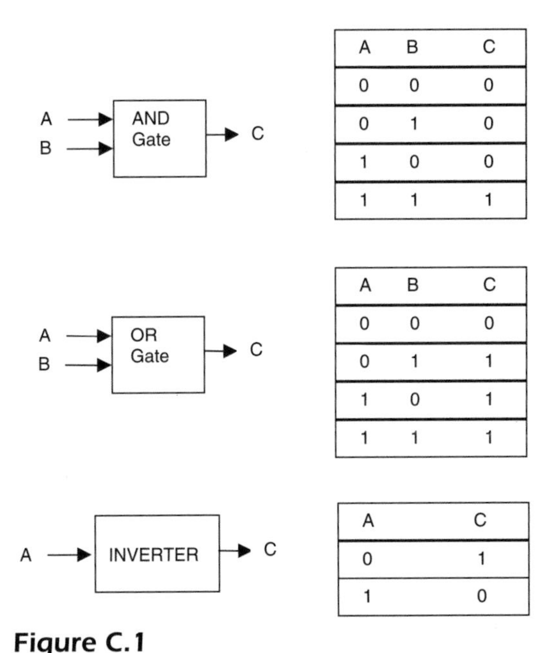

A	B	C
0	0	0
0	1	0
1	0	0
1	1	1

A	B	C
0	0	0
0	1	1
1	0	1
1	1	1

A	C
0	1
1	0

Figure C.1
Logic Gates

Note that:

A logical *AND* gate's output is a "1" only if **all** inputs are "1"s.

The output of a logical *OR gate* is a "1" if **any** input is a "1".

The output of an *inverter* gate is a "0" if the input is a "1".

The different gates are often represented with different symbols, other than the box symbols shown here.

There are many combinations and variations of these basic three functions (e.g., NAND, NOR, or EXCLUSIVE-OR). We can implement **any** logical function, no matter how complex, using just these few gates.

Logical operations include *and, or, compare, greater than, less than, add, subtract*, and so forth. All complex mathematical calculations are built on these simple logical operations. That is how a computer is able to calculate.

Did You Know?

The Apollo spacecraft guidance computer was built entirely of a single type of gate. Only one kind of integrated circuit needed to be developed and exhaustively tested. This approach successfully ensured the reliability of the computer on the flights to the moon and back.

SIGNAL DELAY

The inputs and outputs of gates are stable in only two *steady states:* a "1" or a "0". However, when they change from a "1" to a "0", they are in *transition.* It takes a finite time for the signal to rise or fall between "0" to a "1". This transition time is called *rise time* or *fall time.*

It also takes time for a signal change to travel (*propagate*) through a gate or along a wire. This time is called the *propagation delay.*

The *driving gate* output signal goes to inputs at one or more *receiving gates.* The *delay time* depends on the wire length, resistance, and capacitance. It also depends on the number of receiving gate inputs. A low resistance driver transistor or *buffer* can reduce the *signal delay.*

COMPUTERS

We said that logical gates can do any kind of mathematical operation. A *processor* is a block of logic gates, registers, and control logic. The processor can perform complex operations—add, subtract, multiply, divide, compare, etc. Each operation is controlled by an *instruction.* The processor is often called a *central processing unit (CPU).*

A *computer* is a combination of a processor, memory, and instructions stored in the memory. The computer executes a sequence of instructions to perform a useful task. Each instruction is relatively small and simple, but can be executed very rapidly.

Some computers have many different processor and memory ICs mounted on PC boards. Other computers can fit on a single IC chip and are called *microprocessors* or *microcomputers.* In addition, some chips can hold multiple processors and multiple blocks of memory.

SOFTWARE

Software is the sequence of instructions that tell the computer what to do. Each complete software sequence is called a *program.* (Short chunks of programs are called *subroutines.*) There are programs for all kinds of applications. Examples include word processing, spreadsheets, presentations, payroll, and graphic design.

Programmers write the application programs. Various high-level *software languages* help programmers create the instructions. The languages are specialized to fit the application areas.

Electronic Design Automation programs form an entire class of applications. EDA has languages to help define and create IC designs and test patterns.

Appendix D

Metrics

In this appendix...

- Introduction
- Number Abbreviations
- Time Measures
- Semiconductor Related Sizes

INTRODUCTION

EDA and semiconductor technology jargon is a confusing mixture of English and metric terms and measures. The reader is likely to run into many of these terms and dimensions. Therefore, we provide three quick reference tables to help with unfamiliar names and sizes.

Table D.1 provides a list of common number terms. Two columns show the terms for large and small numbers in long and short (exponent) forms. Two other columns give the English and metric terms, where they exist.

Table D.2 lists the names of time-related numbers. The example column gives some examples so the reader can gain a sense of relative size to familiar things.

Table D.3 describes relative sizes of items in the EDA and semiconductor world (in increasing order). Many sizes are often expressed in both metric and English terms, so we give some equivalent values.

Small and large numbers of objects are difficult for most people to visualize. Here are some real-world comparisons that might help.

Small Numbers

A "mil" is one-thousandth of an inch. Most people are familiar with plastic sandwich or garbage bags. The garbage bag thickness is about one mil (and usually labeled as such). A human hair is one to three mils in diameter.

A meter is about three inches longer than a yard. A "mm" is one-thousandth of a meter (millimeter). A millimeter is about 39 mils. Most people are familiar with the little round black (poppy) seeds sprinkled on bread, rolls, and bagels in grocery stores and bakeries. They are just about one mm in diameter.

A micron is one-millionth of a meter, one-thousandth of a millimeter. This is the hardest to visualize, since it is almost invisible. It is about one twenty-fifth smaller than a fine human hair. Spider web filaments are two or three microns. There are about 25 microns to a mil.

Large Numbers

Visualizing large numbers is also difficult. Here is one way to visualize one thousand, one million, or one billion of something.

Imagine you have a small table (card table) about one yard or meter square. If you line up a row of our little poppy seeds along one edge the table, you will see about one thousand seeds.

If you cover the tabletop with the seeds, you will see about one million (mega) seeds.

Now put a box on the table that is about the same size square and deep (one meter wide by one meter long by one meter deep) as the table. Imagine filling it with seeds. You would see 1,000 million (giga) seeds.

Table D.1 Number Abbreviations

NUMBER	EXPONENT FORM	ENGLISH	METRIC
1	10^0	One	
10	10^1	Ten	Deka
100	10^2	Hundred	Centa
1000	10^3	Thousand	Kilo (K)
1,000,000	10^6	Million	Mega (M)
1,000,000,000	10^9	Billion	Giga (G)
1,000,000,000,000	10^{12}	1000 Billion	Tera (T)
1/ 10th	10^{-1}	Tenth	deci (d)
1/100th	10^{-2}	Hundredth	centi (c)
1/1000th	10^{-3}	Thousandth	Milli (m)
1/1,000,000th	10^{-6}	Millionth	Micro
1/1,000,000,000th	10^{-9}	Billionth	Nano (n)
1/1,000,000,000,000	10^{-12}	Trillionth	Pico (p)

Table D.2 Time Measures (reference: Scientific American 9/2002)

NAME	10?	DESCRIPTION	EXAMPLE
1 picosecond (psec)	10^{-12}	One thousandth of a billionth of a second	The fastest transistors operate in picoseconds
1 nanosecond (ns)	10^{-9}	One billionth of a second	Light travels about one foot. A PC executes one instruction in 2-4 ns.
1 microsecond (microsec)	10^{-6}	One millionth of a second	Light travels ~ 3 soccer fields
1 millisecond (msec)	10^{-3}	One thousandth of a second	Shortest exposure time of typical camera
1 10th second	10^{-1}	One tenth of a second	Blink of an eye!
1 second (sec)	10^{0}	One second	One heartbeat!
1 minute (min)	60 seconds	One minute	Light from sun reaches earth in ~ 8 minutes
1 hour (hr.)	60×10^{1} minutes	One hour	EDA simulations take many hours
1 day (d.)	24 hrs.	One day	Earth rotates in one day
1 year (yr.)	365 days	One year	Chip designs take from 3 mo. to 2 yr.
1 cycle per sec. (cps, hz)	One hertz	One up and down wave, frequency	Human ear can hear down to ~ 40 hz
60 cps, hz	60 hz	60 hz	U.S. power frequency
3000 hz	3K hz	Three thousand (kilo) hertz	Frequency range of speech 3K-8K hz
15K – 20K hz	15K hz	15 khz	Human ear can hear to about 15-20K hz
100 kHz	100k hz	100,000 hz	Early PC, clock speeds
1 Mhz	10^{6} hz	One million hertz, one megahertz	Later PC, memory clock speeds
1 Ghz	10^{9} hz	One billion hertz, one gigahertz	Recent PC clock speeds
1 Thz	10^{12} hz	One trillion hertz, one terahertz	Fast computer runs 35 tera calculations per sec. (Earth Simulator -2003)

Table D.3 Semiconductor Related Sizes

NAME	DESCRIPTION/ EQUIVALENT	EXAMPLE
Angstrom (A°)	10^{-10} meter, 1/100 millionth of a centimeter	20 A°—Thickness of gate insulator in 0.18 micron transistor process
1 Nanometer (nm)	10^{-9} meter, 0.001 micron	0.18 micron—180 nm. Start of DSM (deep submicron) processes
1 micron (micron)	10^{-6}, one millionth of a meter, one thousandth of a millimeter	Smallest IC feature Analog—2–10 microns Digital—1.0–0.09 micron
1 mil	One thousandth of an inch, 10^{-3} inch	One mil = 25.4 microns Human hair 1-3 mils
1 millimeter (mm)	10^{-3} m, one thousandth of a meter, 1000 microns, ~1/25 (0.040) inch, ~ 39 mils	~ Size of a poppy seed
1 centimeter (cm)	10^{-2} m, one hundredth of a meter, 10 mm, ~ 0.39 inch	Most IC chips are about 1/2–2 cm on a side. (1/4–3/4 in.)
1 inch (in)	1/12th of a foot, one in. = 2.54 cm	Wafer diameters have been 2, 4, 6, 8, 12 inches (~50, 100, 150, 200, 300 mm)
1 foot (ft)	1/3 of a yard, 12 inches, 30.48 cm, 0.30 meters	Large wafer diameter (~300 mm)
1 yard (yd)	3 ft, 36 inches, ~ 0.9 m	English/U.S. length standard
1 meter (m)	One m, one hundred cm, ~ 39 inches	World length standard

Appendix E

References

In this appendix ...

- Conferences
- Organizations
- Standards Groups
- Publications
- EDA Internet Sites
- Universities

The reader may wish to learn more about EDA. This appendix provides useful information sources for that purpose.

Most EDA sources are directed at engineers and EDA programmers. However, overviews, commentary, and marketing-related articles may be of interest to the non-technical reader.

Conferences papers also tend to be quite technical. However, some panels, general session talks, and occasional tutorials may be of broader interest. Some conferences, particularly the Design Automation Conference (DAC), can be both informative and fun for non-technical attendees.

The following list give a sample of some of the larger conferences. A comprehensive calendar of EDA-related conferences is available from the ACM's Special Interest Group on Design Automation (SIGDA): www.sigda.org.

CONFERENCES

DAC, Design Automation Conference, www.dac.com

DAC is the largest EDA and silicon solution event. DAC typically has over 50 technical sessions covering the latest in design methodologies and EDA tool developments. Exhibit and demo suite areas show products from some 200 of the leading EDA, silicon, and IP providers.

The DA conference has an unusual blend of industry and academia. It is perhaps the most interesting conference for non-technical attendees. There are technical sessions, panels, and elaborate exhibits with professional entertainers and large parties. There are canned demonstrations (demos) and serious customer private sessions in closed cubicles. It is usually held in San Diego, Anaheim, or Las Vegas, in late June.

DATE, Design Automation and Test in Europe, www.date-conference.com

This is a major EDA/Test conference in Europe. It is usually held in Europe (Paris or Munich) in February.

DesignCon, www.designcon.com

DesignCon is the first major EDA and system-on-chip (SoC) event each year. It has technical sessions (more industry than academic), exhibits, and product demos. It is usually held in San Jose in January or February.

ESC, Embedded Systems Conference, www.esconline.com

The ESC holds several conferences a year in the U.S. (East, West and Central), Europe, and Japan. It focuses on embedded systems, real-time products and tools, with papers and tutorials.

ICCAD, International Conference on Computer-Aided Design, www.iccad.com

ICCAD is focuses on information technology for computer-aided design professionals and IC design engineers. This conference is mostly academic, with technical papers and seminars. It is usually held in San Jose in November.

ISQED, International Symposium on Quality in Electronic Design, www.isqed.org

This organization is sponsored by IEEE and has technical papers on broad areas of electronics quality, including EDA. It is usually held in San Jose in the fall.

ISSCC, International Solid-State Circuits Conference, www.isscc.org

The ISSCC presents very technical papers on advances in solid-state circuits and systems-on-a-chip. It is usually held in February in the U.S.

ITC, International Test Conference, www.itctestweek.org

This is a major test and test equipment conference. ITC covers electronic test of devices, boards, and systems. It is usually attended by academia, design tool and equipment suppliers, designers, and test engineers. It is held in September or October, usually in the U.S.

ORGANIZATIONS

EDAC, Electronic Design Automation Consortium, www.edac.org

EDAC works on business issues relevant to EDA companies worldwide. It is a good source of EDA industry statistics. It gives the Kaufman award yearly to major EDA contributors.

FSA, Fabless Semiconductor Association, www.fsa.org

FSA fosters trade between FSA partners. These include fabless semiconductor companies, wafer foundries, packaging and test companies, investment bankers, venture capitalists, consultants, and intellectual property providers.

IEEE, Institute of Electrical and Electronic Engineers, www.ieee.org

The IEEE is a non-profit, technical professional association of thousands of individual members in 150 countries. It has developed hundreds of consensus-based standards. IEEE sponsors many conferences and publishes innumerable technical papers.

IEEE-CAS, IEEE Circuits And Systems Society, www.ieee-cas.org

This is the IEEE group concerned with EDA theory, analysis, and design. It is mostly academic, with some industry participation.

ITRS, International Technology Roadmap for Semiconductors, www.public.itrs.net

The ITRS identifies the technological challenges and needs facing the semiconductor industry over the next 15 years.

It is sponsored by the Semiconductor Industry Association (SIA), the European Electronic Component Association (EECA), the Japan Electronics & Information Technology Industries Association (JEITA), the Korean Semiconductor Industry Association (KSIA), and Taiwan Semiconductor Industry Association (TSIA).

SEMI, Semiconductor Equipment and Materials International, www.semi.org

SEMI represents some 2,400 companies serving the global semiconductor equipment, materials, and flat panel display industries.

SEMATECH, www.sematech.org

This is an international industry consortium of semiconductor-related companies. SEMATECH develops near-term advances in semiconductor technology.

SIGDA, Special Interest Group on Design Automation of the ACM (Association of Computing Machines), www.sigda.org

SIGDA is the ACM technical society for EDA professionals involved in the application of the computer to all phases of electrical and electronic design. (It co-sponsors DAC, DATE, ICCAD, and other leading conferences, workshops, and other projects.) (ACM is a software organization, despite its name.)

SI2, Silicon Integration Initiative, Inc., www.si2.org

SI2 is an open source EDA standards consortium involved in several standards efforts.

WSTS, World Trade Semiconductor Trade Statistics, www.wsts.org

This is a non-profit group with statistics on semiconductor sales worldwide.

STANDARDS GROUPS ....

Accellera, www.accellera.org

Accellera is a consortium of EDA vendors formed by the unification of Open Verilog International and VHDL International. It promotes a language-based design automation process.

AEA, American Electronics Association, www.aeanet.org

AEA is the nation's largest high-technology business trade association. Now global with offices in Brussels and Beijing, AEA represents thousands of companies.

EIA, Electronic Industries Alliance, www.eia.org

The EIA represents 80% of the U.S. electronics industry. They promote market development, competitiveness, and standards for their members.

EDAC (see organizations above)

IEC, International Engineering Consortium, www.iec.org

The IEC promotes research and cooperation between industry and academia in the areas of computer science and information technology.

IEEE (see organizations above)

OAC, Open-Access Coalition, www.si2.org

Led by Silicon Integration Initiative (SI2), this user-backed group has established an industry-wide data model and application programming interface (API) to potentially access any database.

OCP, Open Core Protocol, www.sonicsinc.com

OCP is an international partnership to standardize an efficient core connection standard to facilitate plug-and-play SoC design. (It has been adopted by UMC, MIPS, Nokia, Texas Instruments, Sonics, and others.)

OSCI, Open Systems C Initiative, www.systemc.org

OSCI promotes SystemC as a de facto open source standard for system-level design. SystemC is a design and verification language built in C++ that spans from concept to implementation in hardware and software.

VSIA, Virtual Socket Interface Alliance, www.vsi.org

VSIA provides open standards, specifications, and guidelines for the mix and match of Virtual Components (IP) from multiple sources.

X-Initiative, www.xinitiative.org

The mission of the X Initiative is to accelerate the readiness of the supply chain to fabricate X Architecture chips (45-degree angle wiring).

PUBLICATIONS

EE Times, Electronic Engineering Times, www.eetimes.com

This is a weekly newsmagazine for design and development engineers and technical managers. It covers EDA developments as well as electronic design. Print and on-line versions, eetimes.com. Related international publications are E*E Times Asia, EE Times China, EE Times UK.*

Electronic Business, www.eb-mag.com

This is an executive-level business magazine with specific focus on the electronics industry for senior electronics managers.

Electronic Design, www.elecdesign.com

Electronic Design covers electronic design bi-weekly for a global audience. It has a TechView section on EDA.

EDN, Electronic Design News, www.ednmag.com

This is a biweekly magazine for design engineers and engineering managers worldwide (EDN, *EDN Europe*, *EDN Asia*, *EDN China*, *EDN Japan*, and *EDN Access*).

ACM Transactions on Design Automation of Electronic Systems (TODAES), www.acm.org/pubs/contents/journals/todaes

This is a journal of EDA research work, published by the ACM (Association of Computing Machinery) which is the leading technical organization for software programmers, despite its name.

EDA INTERNET SITES

www.edatoolcafe.com Weekly update on EDA news

www.eetimesnetwork.com Weekly update on IC and EDA news

www.edtn.com Electronics Design Technology and News Network

www.edtn.com/encyclopedia This site has a guide to technical terminology

www.e-insite.net This website has a large acronyms and abbreviations list.

http://internet.about.com/cs/softwareeda This has a link to recommended EDA-related books

www.sematech.org/public/publications/dict/index.htm This site has a guide to technical terminology

www.deepchip.com Synopsys Users Group (SNUG) This site has an EDA tool commentary by users.

http://www.synopsys.com/corporate/invest/EDA_Primer.pdf This gives a brief overview of EDA for financial analysts

UNIVERSITIES

Since readers have different interests (research programs, professors, courses, etc.) we list only a sample of leading U.S. schools. There is much work going on in Europe and Asia as well.

Website locations of specific EDA areas of interest may be found under different prefixes or suffixes. These include Electrical Engineering (ee), Computer Engineering (ce), Computer Science (cs), ece, eecs, and the like.

Sometimes the EDA area is under VLSI Design, Computer Aided Design (CAD), Computer Aided Engineering (CAE), Design Automation (DA), or EDA. Some universities have "Centers" for related subjects, with EDA research included. Sometimes they are listed under the specific professor's name.

One can also identify schools with EDA interest from conference program committee contacts. EDA, ASIC, and FPGA companies also sponsor active EDA research programs.

A partial list of universities with significant EDA programs, in alphabetical order, includes:

Caltech—California Institute of Technology, www.cs.caltech.edu

CMU—Carnegie Mellon University, www.cmu.edu, www.cs.cmu.edu

Georgia Tech—Georgia Institute of Technology, www.ece.gatech.edu

MIT—Massachusetts Institute of Technology, www.mit.edu

PU—Princeton University, www.princeton.edu

Stanford—Stanford University, www.stanford.edu

UCB—University of California at Berkeley, www.berkeley.edu, www.cad.eecs.berkeley.edu

UCLA—University of California at Los Angeles, www.eda.ee.ucla.edu

UCSD—University of California at San Diego, www.ece.ucsd.edu

UIUC—University of Illinois at Urbana, www.ece.uiuc.edu

UT—University of Texas at Austin, www.utexas.edu, Computer Engineering Research Center (CERC), www.cerc.utexas.edu

Appendix F

ICs, IP, and SoC

In this appendix....

THE IC INDUSTRY

An extended hierarchy of companies forms the electronic products *food chain* of suppliers. The manufacturing of silicon chips is one part of this hierarchy.

Figure F.1 shows an overview of this hierarchy (on the left) and how EDA supports it (on the right).

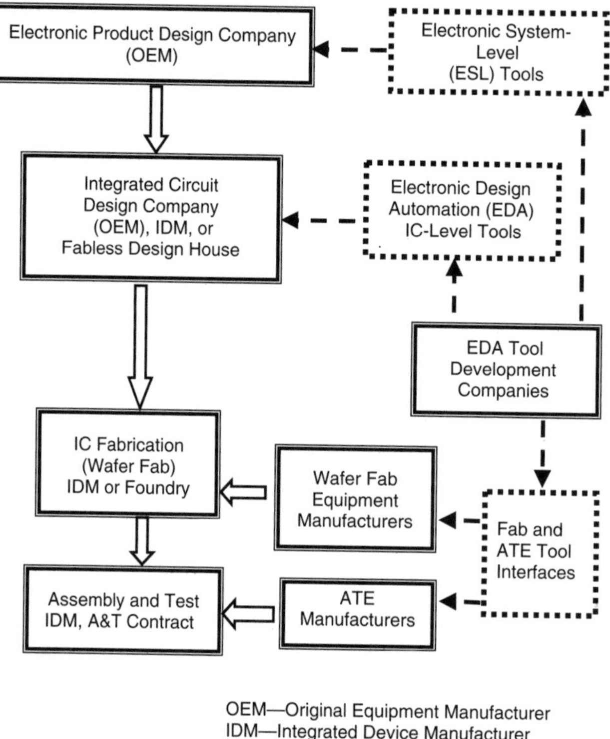

Figure F.1
Electronic Product Hierarchy and EDA

Product Design

Referring to Figure F.1, we see that *end-user* electronic products are made by *Original Equipment Manufacturers* (OEMs). End-user products include communications equipment, video recorders, computers, and cellular telephones. *Electronic System Level (ESL)* EDA tools support system product design (dotted box).

The products have many component parts, some of which are *Integrated Circuits* (ICs). These may be power controllers, microcomputers, memory chips, logic chips, amplifiers, etc. In many products one chip contains **all** the electronics.

Integrated Circuit Design

Design engineers at OEM product companies sometimes design ICs themselves. If not, an *IC manufacturer (IDM)* or a specialty IC design company (*fabless design house*) will do the design. EDA tools support the IC level design work at the OEM, IDM, or fabless design house (dotted box).

The IDM can manufacture the chip with its wafer fabrication plant (*fab*). The fabless design house will contract chip manufacturing with a *foundry* (which does **only** wafer fabrication). The wafer fabs cost billions of dollars and use equipment from *wafer fab equipment* manufacturers.

The IDM may assemble and test the completed chip, or contract out the work. The fabless design house contracts with an *assembly and test* firm to do the work. The test machines cost millions of dollars and use equipment made by *automatic test equipment (ATE)* manufacturers. Both the wafer fab equipment and ATE machines have tool interfaces driven by EDA tools (dotted box).

An IC manufacturer or contract design house usually charges a one-time *non-recurring engineering (NRE)* design fee. The IDM may do the whole IC design with a complete, tested chip for the OEM. If the electronic product company does the complete design, the resultant files are called *customer-owned tooling (COT).* COT designers then own responsibility for the packaging, assembly, test, and manufacturing yield management. They may contract this work out to different contract houses.

The multiple subcontractor approach is known as "*dis-integration.*" Each company focuses on "doing one thing well."

Did You Know?

***Dis-integration* is significant to the EDA industry because the small design houses may have less funding for EDA tools than traditional large OEM and IDM customers. Some of the EDA companies have Design Services groups which compete directly with both the IDMs and the fabless design houses.**

Design Handoff

During the whole design and manufacturing sequence a great deal of information is created. This design information, data, and files are *handed off* (transferred) from group to group. There is lots of room for errors in the design handoff.

Both the "integrated" and "dis-integrated" approaches have their benefits and weaknesses, supporters and detractors. An essential issue with either approach is "Who owns the problem if the chip does not work?"

Historically, the IDM took responsibility for both chip design and manufacturing. With the dis-integration model, there are more players to assume the responsibility (or to say "*not me*!"). The source of failure may also be harder to identify because of all the players involved.

Did You Know?

The integrated/dis-integrated differences are starting to blur. Fabs now cost several billion dollars, so IDMs are becoming partners with each other and are using foundries. And subcontractors are forming teams with foundries to ensure clean handoffs and customer success.

DESIGN RE-USE AND INTELLECTUAL PROPERTY.....

Design Re-use

Teams of engineers strive to put greater functionality on a single chip. Large, complex blocks of prior designs can sometimes be re-used (e.g., a microprocessor, memory, or an Internet access block). This *design re-use* can be extremely valuable to speed up the design effort.

Figure F.2 shows re-use of blocks in a new IC. The design information for a functional block used in an existing chip (A) can be re-used for the same function needed in a new chip (B).

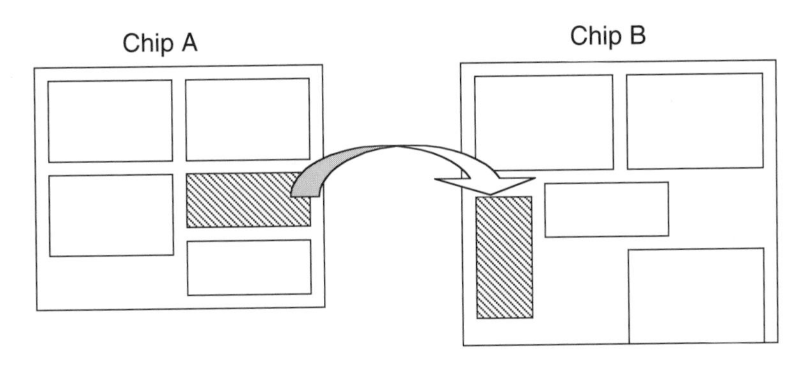

Figure F.2
Design Re-use

Intellectual Property

These complex blocks are complete little designs in their own right. They may be patented or marketed as legal "Intellectual Property" (IP).

Block-based design is a custom chip architecture using pre-designed blocks. There are specialized EDA tools to assist block-based design and to help use and document each kind of block.

Did You Know?

Intellectual Property (IP) **means "re-usable design blocks" in this context. Other names include IP blocks, cores, macros, VCs (Virtual Component) or SIPs (Silicon Intellectual Property).**

When lawyers refer to intellectual property, they use a broader definition that includes trade secrets, formulae, recipes, algorithms, or software programs as well as IP blocks.

In the context of Internet communications, *IP* also refers to *Internet Protocol.* This defines the handshaking messages that computers use to communicate with each other over the Internet. Do not confuse that context with *Intellectual Property* and re-use.

Types of IP Blocks

Soft IP blocks may be pure design information, usually at the RTL design level, independent of any specific semiconductor process. The buyer may choose a library of cells tailored for a specific semiconductor process. The block may or may not have been implemented all the way to silicon (*silicon-proven*). The buyer may be allowed to modify the IP. If so, then the buyer accepts all responsibility for whether or not the IP works.

A *hard IP block* is the final set of mask information targeted to a specific semiconductor library and process. It has been implemented in silicon and is known to work. It will have been characterized (measured for timing, power, area, etc.) by specialized EDA tools. A hard block is more difficult (and risky) for the buyer to modify than soft IP. It also carries less risk since it is proven silicon.

Analog IP blocks are circuit blocks that work with continuous low-level analog signals. *Mixed-signal* circuit blocks convert analog signals into digital numbers, and vice versa. Radio frequency (RF) blocks provide radio receivers and transmitters. Most analog IP blocks are hard IP.

Some FPGA vendors have coined the term *firm IP* for *FPGA IP and Reconfigurable Logic blocks* that can be re-used on their FPGA hardware. An IP block implemented on one FPGA can be re-used on another FPGA (of the same architecture).

The user can set an FPGA chip's on-chip interconnection memory. The same chip's logic can be *re-configured* to perform a different function, which can be swapped in or out of the chip as needed.

Some FPGAs allow very rapid re-configuration, allowing several functions (IP blocks) to be time-shared on one FPGA. These *dynamically reconfigurable* FPGAs can achieve very high speed and low power for selected applications. Some chips used as processors can have the hardware reconfigured for each instruction.

IP Vendor Business Models

Large semiconductor companies often have a database of IP blocks (i.e., all their prior chip designs). In addition, many small companies have formed to market blocks in specialty areas. These areas include memory blocks, standard input/output blocks, processor blocks, and so forth.

There are several kinds of IP business models. Some companies offer a large set of compatible blocks, guaranteed to work together. Other companies market only a specific kind of block (e.g., memories, processors).

Some of the standard, multiple-vendor designs are called *commodity IP*. These are usually standard I/O interfaces used with consumer and industrial electronics. Other leading-edge design blocks, usually from a single supplier, are called *star IP*. They often can demand a higher price than commodity blocks.

Commodity suppliers try to have many customers, but competition keeps the price low. Star IP suppliers may get a higher price, but with fewer customers.

The star IP vendor may have the best-in-class design this year. However, it is very difficult to maintain that leadership place in the "next round" of designs. (And with product lifetimes as low as six to nine months in some markets, the next round comes very quickly.)

Furthermore, some leading-edge users may want exclusive rights to the design. This enables them to maintain an edge over their competition. They may even buy the IP company to get it.

FPGA companies sell *firm IP* designs that can be re-used on their FPGA hardware. Others embed hard blocks like microprocessors into their FPGA arrays. This combines the area density of the hard block with the flexibility of the FPGA.

Some IP vendors sell only hard IP and others only soft IP and some sell both. A few EDA companies also market IP and the EDA tools to support it. Usually they have acquired one or more small IP vendors. Interestingly, they sometimes find themselves in competition with their customers.

The IP industry has grown slowly with mergers, acquisitions, business model struggles, and failures. While the basic idea of IP re-use is sound, there are many issues associated with it.

IP Re-use Issues

Like everything else, there are pros and cons to the re-use of IP blocks. If used without any *rework*, they should shorten design time.

However, beware if a block is modified. If only one line of code or wire is changed, then the entire design block must be re-verified. This may take a very lengthy sequence of simulations and design rule checks. **Any** kind of modification, even removal of an unused section, can be a risky business.

Many companies have desirable IP buried inside a prior product. It was designed for use with that product. Cutting it out and "productizing" it for re-use an another application can be a significant redesign job. Furthermore, it may need to be retargeted to a more recent manufacturing process.

The IP design engineer often does not know in advance how the IP will be used. Therefore, the IP must be thoroughly tested and documented to ensure ease of re-use. The quality (high confidence of successful use) of an IP is a significant purchase issue. Several groups are working to create standard measures of IP quality.

Many people believe that *re-use without rework* is clearly the path of least risk. It should give the fastest design time and the fastest TTM. However, as IP blocks grow more complex, slight variations are frequently desired for different applications.

Several studies showed that in typical chip designs about one-third of the IP blocks are re-used as is. Another third of the blocks are reworked (modified), and about a third of the blocks are entirely new designs.

Another issue is the time (several months) it typically takes for a user company to acquire the desired IP. Sometimes approval from the purchasing and legal departments takes longer than it would to design the block in-house.

Finally there is a legal issue. IP blocks are a form of intellectual property. They can be patented and are subject to patent and copyright infringements. Historically, most IP resided with the large IDMs. There was probably IP infringement on both sides between competitors. Practically, though, they could agree not to sue each other for infringements. (I won't sue you if you don't sue me!) However, this balance might not work with small IP providers competing with large IDMs. Large IDMs typically spend millions of dollars protecting their intellectual property.

SYSTEM-ON-CHIP

With the increase in chip density, we can fit most of an electronic system on a single IC. A chip of this type is called a *System-on-Chip (SoC)*.

Actually, SoCs rarely include the *whole* system. A system may have attached input devices (e.g., keyboard, microphone, or temperature sensor). It may have attached output devices (e.g., display, speakers, or mechanical controller). It needs power sources (e.g., batteries or power converter). But most of the **electronics** may fit on one chip.

A SoC may be just one of the "one or more ICs." Figure F.3 shows the technology migration over time from a rack of equipment to a SoC.

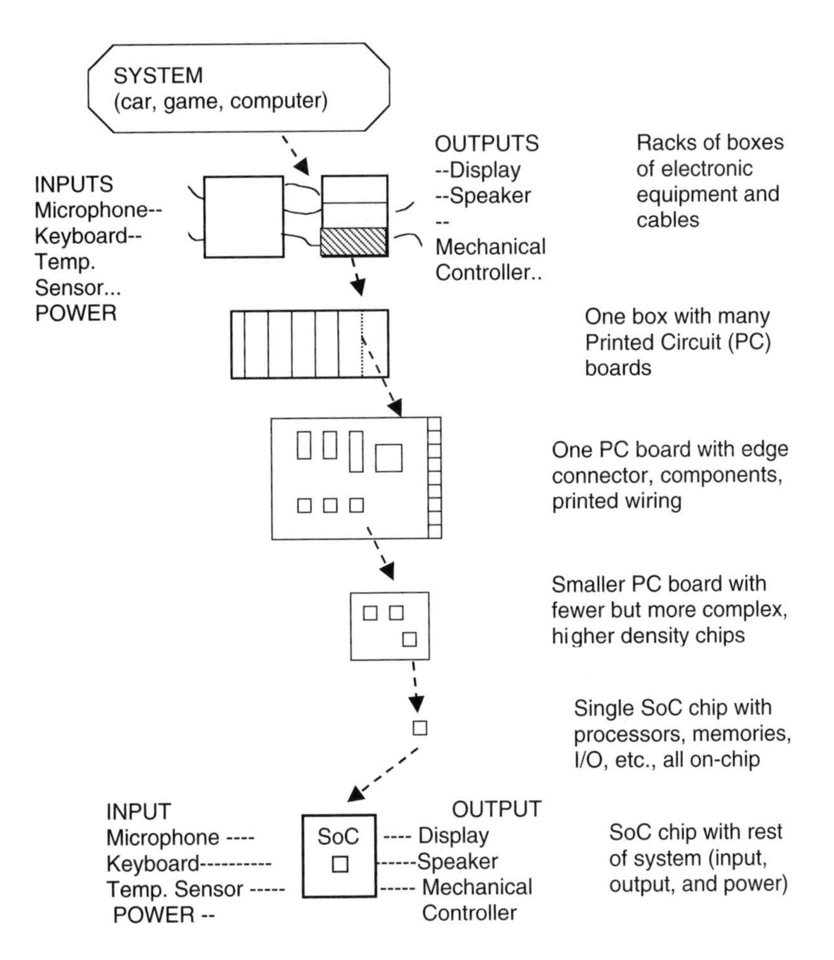

Figure F.3
Migration to System-on-Chip

Looking at Figure F.3 from top to bottom, we see that a system (car, game, or computer) product used to consist of several large boxes of electronics and cables (sometimes *racks* of boxes).

Cables connected inputs (such as microphones, temperature sensors, etc.). Other cables connected outputs (such as displays or speakers). The electronics was on PC boards that plugged into the boxes.

On the PC boards were many electronic components. As more components were combined into ICs, the PC boards got smaller. Finally, all the components were combined into one IC—the **system-on-chip.**

The bottom picture on Figure F.3 shows the SoC electronics chip with the input and output connections needed to complete the system. The SoC chip usually mounts on a small PC board in order to connect to everything else.

Often a SoC will include one or more processors, memory blocks, and I/O (input /output) blocks. It may have several IP blocks, and often embedded firmware or software.

Did You Know?

Though some attempts have been made, there is no industry agreement on what constitutes a SoC. (Nor does there appear to be a pressing need for a rigorous definition!)

SoC Issues

The chip architectures for logic, programmable, analog, mixed signal and memory each use different methodologies. They use different chip processes, voltages, and power distribution. These differences make it a difficult task to merge them on a single chip.

Each architecture also uses different EDA tools. The design results of these tools must also be merged to create the integrated design. For some applications, multiple separate chips may be simpler and more cost-effective (for example, RF or analog front-end, digital control, memory, and power driver I/O chips). Sometimes these are assembled or stacked into one package called a system-in-package (SIP). (This is not the same SIP as the IP block mentioned earlier.)

The SoC idea is exciting, but there are many issues with it. There are several conflicting requirements. There is a huge consumer market for low-cost portable electronics of all kinds. However, the high-volume consumer market products typically have short lives (often less than one year). Consumer markets thus require a shorter time from idea to market than traditional IC markets.

One dilemma is that SoCs take longer to design. The increased complexity requires more work and the skills of larger teams of engineers. The latest technology also increases the design time with more design constraints, tests, and checks. Long wires on large chips slow the signals, making it harder to achieve performance goals. In addition, a large complex chip takes longer to test, further increasing the cost.

Another conflict is that portable consumer products use batteries, which need low-power chips. More functions demand higher chip speeds. However, big, fast chips usually require more power, not less.

Often, designers needed two or more silicon passes (*re-spins*) to get a working chip. However, the short time to market for typical SoCs requires a shorter design time and first-pass silicon success. The high cost of mask sets also presents a major barrier to re-spin a chip.

In addition, legal and business issues come with third-party IP. And while good internal company IP may be embedded in an older product, it may be difficult to extract.

Platforms

Platforms are a form of *SoC re-use* intended to speed up the TTM. The platform concept raises the "re-use" idea to the next level of abstraction above blocks.

A pre-designed "platform" is targeted (functions, power, performance, cost, etc.) for a specific market area. Some portions of the design are intended to be modified, tailored, or customized by the customer. The intent is to enable a differentiated product to get quickly to market.

Did You Know?

Platform is another term with evolving definitions. The idea is to assemble a group of compatible blocks for a specific market area (e.g., game controls, cellular telephone, etc.). The blocks are all pre-tested to ensure that they work together (interoperability).

Platform architectures can use multiple standard cell or custom IP blocks with one or more configurable blocks. However, the configurable FPGA blocks tend to be much larger than the custom IP blocks. They require different power and routing approaches. Another tough question is predicting **where** the user will want to make changes. Where are the best places to add the configurable blocks?

A competing alternative is to start with an FPGA base architecture with other custom blocks embedded. This ensures configurable material all over the chip. Again, the difficult questions are how much configurability is needed and where will it be needed?

Both of these approaches have significant EDA tool implications. And each approach naturally has its supporters and detractors.

SUMMARY

The electronic product and IC industries are both supported by the EDA industry. EDA supplies tools to electronic system product design and IC design companies. The EDA companies also support wafer fab equipment and automatic test equipment manufacturers.

Design re-use can significantly speed up complex IC designs. IP vendors form a slowly growing industry.

Many issues pro and con surround the use of IP and the difficulties of re-use in general.

System-on-chip design is also growing. It has many implementation problems, however. At a higher level, the use of platforms is another SoC approach to rapid TTM.

Appendix G

EDA Glossary—Terms and Acronyms

Note: For acronym pronunciation, the default case is to simply spell out the letters. Otherwise, the pronunciation is given in *italics* within parentheses, e.g., ASIC *(A-sick).*

Term or Acronym	Definition
Abstraction	Levels of details—higher levels of abstraction hide lower levels of details. (See also **Hierarchy** and **Design views**.)
ACM	Association of Computing Machinery (programmer's association).
Accelerators	Usually a hardware device to speed up a simulation or modeling program. Either a card or a special computer.
Accellera	Standards group for a formal verification language, as well as VHDL and Verilog.
Access	The general requirement to reach all the logic inputs during test.
A/D	Analog-to-digital conversion *(A-to-D)* converts analog signals to digital numbers.
ADC	Analog-to-Digital converter. (See also **A/D**.)
ADSL	Asymmetric digital subscriber line—a form of telephone transmission to handle both voice and data (at different rates).
AC, alternating current	Current that alternately flows back and forth at a regular rate.
AEA	American Electronics Association.
ALU	Arithmetic and logical unit, core block in computers.
All-angle routing	Interconnect wiring defined at all angles (often used in analog circuits).
AM	Amplitude modulation. Radio waves varied in height (amplitude).
Amperes	Measure of the amount of electrical current (also milliamps, microamps).
Amplify	Using a small voltage or current to control a large voltage or current.
Analog	Signals which vary smoothly and continuously (in contrast to digital).

AND gate	Logic gate that gives a "1" output only if all inputs are "1s". *(and)*
Antenna effects	Layout patterns that cause wire stubs affecting signal integrity.
API	Application programming interface.
ARM	Advanced RISC machine (and microprocessor company). *(arm)*
ARPA	Advanced Research Projects Agency (also DARPA). U.S. Dept. of Defense research funding agency. *(ar-pa, dar-pa)*
ASCII	American Standard Code for Information Interchange—basic text format. *(ask-y)*
Assembler, assembly code, assembly language	Software that translates from human-readable low-level assembly language into machine code (1s and 0s).
Assembly and test	Refers to the steps of assembling the chip into a package and testing the final packaged chip.
Assertions	Statements entered for formal verification of a logic design.
ASIC	Application-specific integrated circuit. Usually semi-custom architecture (gate arrays or standard cells) to reduce design time. *(A-sick)*
ASIC Council	Group of leading ASIC manufacturers.
ASP	Application-specific product, average sales price, application service provider business model.
ASSP	Application Specific Standard Product.
Asynchronous	Any signal lacking a regular, predictable timing relationship. The output state of a circuit may be independent of the clock signal, and the operational speed of a signal depends only on its propagation delay through a network, rather than on clock pulses.
ATE	Automatic Test Equipment.
ATPG	Automatic Test Pattern Generation. ATPG is generally used to augment functional test patterns to increase the level of fault coverage.
Back-annotation	The process of updating simulation files with delays extracted from the actual post-layout wiring.
Back-end	Back-end design—physical design steps of IC design.
Backward compatibility	When an upgrade of a device or software will run with things that worked with the prior device or software.
Bandwidth	Measure of a circuit's ability to process or transmit information.

Behavioral description A model of a device or function in terms of algorithms or mathematical equations, both function and timing.

Benchmark A standard test by which products are evaluated and compared.

BGA Ball-Grid Array.

Bidirectional A signal or port that can act as either an input or output.

BiCMOS A mixed-technology process generally used to fabricate ASICs consisting of bipolar I/O cells and a CMOS core. The bipolar I/O transistors provide fast switching and high-output drive, while the CMOS transistors comprising the core cells provide high functional density and also conserve power. *(bi-C-mos)*

Binary Digital signals or symbols which occur only at two levels (e.g., high or low, 0 or 1, on or off). Binary number system—based on two, with just two symbols.

Bipolar A process technology employing two junction transistors: NPN and PNP.

BIST Built-In Self-Test. A method of generating test vectors on chip and summarizing outputs in a signature register, to rapidly test an IC at speed without external equipment. *(bist)*

Bit Smallest unit of binary data (1 or 0, H or L, etc.). (See also **Byte**.)

Black box Implementing or testing an IP block whose internal details are hidden from the designer.

Block-based design A design style that builds on pre-designed functional blocks.

Block diagram A design picture that represents functions and components as blocks.

Bonding wires Fine wires that connect the contact pads on the IC to the IC package pins.

Boolean algebra, logic Branch of math to describe logic. (See also **Logic**.)

Bottom-up design A design style that begins with the lowest-level blocks and builds the design upward from there. (See also **Middle-out** and **Top-down**.)

Boundary scan A method of inserting registers and control logic in an IC to scan test patterns in serially, and scan output data registers, to test an IC.

Breadboard A printed circuit board suitable for building experimental circuits.

Buffer, buffer circuit	A low-impedance inverting driver circuit that can supply substantially more output than the basic circuit. The buffer element is used for driving heavily loaded circuits or minimizing rise-time deterioration due to capacitive loading.
Bugs	Errors or problems in a design or device.
Burn-in	The process of running a chip or system at full power for an extended time period to weed out any early failures.
Byte	Eight-bit chunk of data. (See also **Word**.)
C	Popular programming language.
CAD	Computer-Aided Design. (See also **EDA**, **ECAD**.) *(C-A-D or cad)*
CAE	Computer-aided engineering.
CAGR	Compound annual growth rate. *(cag-er)*
CAM	Computer-aided manufacturing. *(cam)*
Capacitance	Measure of amount of storage which a device has for electrons.
CBA	Cell-based array.
Cell	A predefined layout of circuit elements that implements a specific electrical function.
Cell library	A collection of cells whose characteristics are generally specific to an ASIC vendor.
Channel router	EDA tool to define wiring paths in channels between cells.
Characterization	The statistical testing of a chip or library to determine the usable range of the many parameters (voltage, current, speed, etc.).
Chip	Alternate name for integrated circuit.
Chip scale package	A small IC package not much larger than the IC chip itself.
CIC	Customer Identification Code.
Circuit	Electrical circuit—functional group of electronic components and their interconnect wires.
Circuit design	Creation of a functional grouping of components by selection of the electrical components and their interconnection.
CIS	Component Information System—standard supply chain component information system. Also component supplier management (CSM).
CISC	Complex Instruction Set Computer (vs. RISC—reduced instruction set computer). *(cisk)*

Clock(s) Timing pulses which control the sequences and timing of each control and data operation in the logic.

Clock skew The phase shift in a single clock distribution network resulting from the different delays in clock-driving elements and/or different distribution paths.

Clock tree A clock distribution technique that minimizes clock skew.

CMOS Complementary metal-oxide semiconductor. (See also **MOS**.) *(C-mos)*

CMP Chemical mechanical polishing.

COL Customer-owned layout.

Commodity IP Standard IP blocks made by multiple vendors. (See also **Star IP**.)

Compiler A programming tool that translates a high-level language down to machine language (object code).

Conductor A material such as aluminum or copper that allows electrons to flow easily.

Constraints Any set of rules, barriers, or restrictions that limits the variability of a design.

Control logic The part of logic design that describes how and when something gets done, such as the transfer of data between registers.

Copper wiring Wafer manufacture using copper instead of aluminum for wiring.

Core The active or used area of an ASIC or the area excluding the I/O pad ring.

COT Customer-owned tooling.

CPU Central processing unit. (See also **Processor**.)

CPLD Complex programmable logic device.

Crosstalk, coupling The influence of a signal in one wire on a signal in another wire.

CTS Clock tree synthesis.

Current sink The place current returns to in an electrical circuit (such as the negative terminal of a battery).

Current source The place current starts from in an electrical circuit (such as the positive terminal of a battery).

Custom IC IC in which all components are designed just for that IC (no re-use of prior designs). (See also **Full Custom**.)

Cycle-based simulator Simulator that updates values at each computer instruction cycle. It is *cycle-accurate* if the timing is correct within the instruction cycle.

DAC Design Automation Conference (major EDA conference). *(dac)*

Also, digital-to-analog converter.

Data Information stored or transferred between electronic components (e.g., numbers, letters, voice, or video information).

Data path A bus-oriented circuit architecture optimized for pipelined or bit-parallel operations. Data paths typically include data processing elements such as ALUs, muxes, shift registers, and other register-transfer functions.

Data representation The way the data (numbers, units, coding, etc.) is represented in a particular data file. Also, data structure.

DATE Design Automation and Test, Europe conference. *(date)*

DC, direct current Current that flows in one direction only (as from a battery).

DCL Delay calculation language.

Debug, debugger The fixing or removal of bugs (problems). A program tool to help debug software.

Decimal system Number system based on ten, with ten symbols, zero through nine.

Defect A fault condition resulting from contamination or a manufacturing anomaly that may or may not affect circuit operation.

Defect density The number of defects statistically present in a given area of the wafer. A typical defect density for a process in a class ten or better clean room is one to two defects per square centimeter of wafer area.

Delay fault See **Race condition**.

Delay model(s) Delay calculators use mathematical models of devices and wires to compute the delay through the devices and wires.

DES U.S. government-sponsored data encryption standard. *(des)*

Design capture, entry Entering the designer's ideas (text or symbols) into the computer.

Design closure Refers to the process of finding and fixing all design errors. (See also **Timing closure**.)

DesignCon EDA and SoC conference.

Design exploration Investigation of different design approaches and architectures.

Design files	The collection of data files which constitute all the information about the hardware or software being designed.
Design flow	The sequence of specific tools and files required to design something (such as an IC).
Design guidelines	Various things the designer should do to create a successful design.
Design methodology	The sequence of design steps necessary to design something.
Design redundancy	Technique of adding duplicate circuits to improve reliability.
Design repair	Automatic correction of design rule error by an EDA tool.
Design re-use	The use of prior design blocks as part of a new design.
Design rules, design rule check	The recipe for a process. Design rules specify minimum width and spacing requirements for the polygons comprising the physical layout.
Design space	The area the designer can work in, defined by the constraints on size, power, voltage, cost, performance, etc.
Design trade-offs	Set of design variables or factors (e.g., cost, performance, power) which the designer must balance (trade off) during the initial design stages.
Design views	The various representations of a design—physical, logical, behavioral, timing, etc.
Detail routing	Defining the interconnect paths on a chip within blocks.
Device	Any electrical component. Also, the transistors, resistors, and capacitors in an IC.
Device library	Electrical and physical device data for a group of devices (often used during synthesis).
Device modeling	Software models of the IC devices based on process parameters.
DFM	Design for manufacturing—including manufacturing constraints in the design process.
DFT	Design for test—including test constraints in the design process.
Diagnostic test	A test which helps identify or isolate the cause of an error.
Die, dice	Also called a chip, a die is an individual circuit sawn or broken from a wafer that contains an array of such circuits or devices.
Die attach	The process of attaching the die to the package cavity or substrate. Die attach materials include epoxy, gold, and silver-filled glass.

Diffuse, diffusion	Embedding atoms of one material into another to form a mix or alloy, doping.
Digital	Signals which occur only in two or more discrete levels (in contrast to analog).
DIP	Dual-in-line package. *(dip)*
Dis-integration	The migration from the integrated device manufacturer to dis-integrated firms which each do part of the IC design and manufacture.
Distributed design	Design work on the same project, by different groups, usually geographically separated.
Doping	See **Diffusion**, **Implant**.
Drain	Current sink in a MOS transistor.
DRAM	Dynamic random access memory. (See also **Memory**.) *(D-ram)*
DRC	Design rule checker, a tool that checks an IC design for physical design rules violations, e.g., spacings, thicknesses, etc.
DSM	Deep sub-micron (<0.18 minimum feature size process).
DSP	Digital signal processor or processing, a type of processing used in communications systems.
DTA	Dynamic timing analyzer. A program that measures logic timing by simulating the logic operation with timing information. Test patterns are needed to drive the simulation.
Dynamic	Any type of device testing during which the clock is applied.
	A memory element in which logic state storage depends on capacitively charged circuit elements. These elements must be continually refreshed or recharged at regular intervals.
Dynamic re-configuration	The ability to rapidly change the programming of all or portions of an FPGA memory.
EBL	Electron beam lithography—used to make fine dimension masks.
ECAD	Electronic computer-aided design. *(E-cad)*
ECO	Engineering change order (formal change control instructions).
EDA	Electronic design automation (umbrella term for CAD, CAE), software tools to aid the design of electronic systems and chips.
EDAC	Electronic Design Automation Consortium. *(E-dac)*
EDIF	Electronic design interchange format. Early standard for exchange of IC design data between manufacturers. *(E-dif)*

Editors	Program which enables entry or revision of text or symbols during design capture (hardware or software).
EE Times	Electronic Engineering Times—weekly newsmagazine.
EEPROM	Electrically erasable programmable read-only memory. (See also **Memory**.) *(E-E-prom)*
EIA	Electronics Industry Association.
Electronic Business	Magazine.
Electronic Design	Magazine.
Electronic Design News	Magazine.
Embedded system	Microcomputer embedded in control system, usually non-accessible to user.
Embedded software	Software that runs on an embedded system, often small, robust, in real time.
EMC, EMI	Electro-magnetic compatibility, interference. High-frequency noise generated by one device which can affect others.
Emulation	The process in which a device being developed is prototyped prior to manufacture.
EPROM	Electrically programmable read-only memory. (See also **Memory**.) *(E-prom)*
Equivalence checker	A form of formal verification which checks to make sure that two logic descriptions are equivalent (such as before and after synthesis).
ERC	Electrical rule checker; a program that checks a circuit for electrical rules violations such as excessive fan-out, opens, and shorts.
ESC	Embedded System Conference.
ESD	Electrostatic discharge. When electrostatic-induced voltages are discharged at a device's input pins, physical damage to the device is likely to result (*aka* zap, zapping).
ESDA, ESL	Electronic System Design Automation, *aka* Electronic System-Level Design (ESL) design.
Event-based simulation	Simulator keeps track of events such as only when each gate switches. Also, event-driven simulation.
Exclusive OR gate	Logic gate that gives a "1" output if only one input is a "1".

Exhaustive test	Testing that checks all possible combinations of input variables. Often not possible within real-world time or cost budgets.
Extraction	The process of calculating the net lengths, resistances, capacitances, etc., from the actual physical design.
fab	Fabrication—the semiconductor fabrication. (See also **Foundry**.) *(fab)*
Fabless design house	Company that designs ICs, but without internal wafer fab.
Farads	Measure of capacitance in an electrical circuit (also microfarads, picofarads).
Fault, stuck-at	A manufacturing failure which causes a signal to be stuck high or low.
Fault coverage	Refers to the percentage of possible fault conditions that can be detected by a production test program.
Fault grading	A computer-automated process of determining fault coverage by simulating faults that are modeled as circuit nodes which are stuck at a logic 0 (open) or a logic 1 (short).
Farm, server farm	Group of computers sharing storage access and compute load.
Feature size	The size of the parts of a transistor. The smallest feature, usually the gate width, is used to label the process that produces it (i.e., 0.13 micron is both the smallest feature size and the name of the process technology).
Feedthrough	See **Via**.
FFT	Fast Fourier transform.
FIFO	First in, first out. *(fife-O)*
File format	The detailed way the data is arranged in a data file.
fingers	Printed copper contact connections for PC boards.
Finite element modeling	Numerical method used for complex models composed of many tiny elements.
First silicon	Critical step for a new IC design—the first silicon chips made.
Flat design	A flat design shows the finest level of detail. Any hierarchical view has been removed or "flattened." (See also **Abstraction** and **Hierarchy**.)
Flip chips	IC chips with bonding pads made to allow the chips to be bonded face down to a package or PC board.
Flip-flops	A logical circuit with memory that can be set to a 1 or 0.

Floating license An EDA tool license which can be applied to more than one workstation or "seat."

Floorplanning The process of placing functional blocks within the chip layout and allocating interconnect routing between them such that an optimum layout is achieved.

Floating Node A gate input or output that is erroneously left unconnected, resulting in functional failure. Floating nodes will generally float to a logic high state.

FM Frequency modulation. Radio waves frequency varied to communicate information.

Food chain General term to describe a hierarchy of elements where each supports or contributes to the one(s) above it.

Footprint The physical outline of a logic block on an IC.

Formal verification Verification using formal mathematics to "prove" the correctness of a logic design. Designers enter "assertions" and "properties" that describe the intent of the logical function.

Foundry A semiconductor fabrication facility that makes its excess manufacturing capacity available to other semiconductor companies or customers with their own process-compatible tooling.

FPGA Field programmable gate array. An IC architecture of fixed arrays of complex gates and a fixed mesh of interconnect wires. The interconnections may be user-"programmed" by RAM or Flash memory, or by fusible links.

Frameworks A CAE environment that provides a common design database and user interface, allowing the transfer of design data from one tool to the next without the risk of translation errors.

Front-end design Front-end steps of IC design, e.g., design capture, verification, and synthesis.

FSA Fabless Semiconductor Association. Trade group of IC foundries, design houses, and related support companies.

FSM Finite state machine. A structured way to design control logic.

Full-custom An integrated circuit design in which each circuit element is individually drawn and positioned in the chip layout. All mask layers are specific to the design.

Function, functional Refers to the logical operation of a design, without timing. With time descriptions, the operation or modeling is behavioral.

GA Gate array—an IC architecture consisting of a fixed array of standard gates. The customer defines the interconnections. (*gate array*)

GaAs A non-silicon process technology characterized by ultrafast operating frequencies. *(gallium arsenide)*

Gate A circuit having two or more inputs and a single output, the output state being a function of the combination of logic signals at the inputs. The fundamental logic gate types perform Boolean functions such as AND, OR, NAND, and NOR.

Gate count A metric for the size of an IC design.

Gate array A semicustom ASIC technology that utilizes prefabricated wafers processed up to the final metal interconnect layers. These generic wafers are then customized as a function of the user's design, which defines the connectivity of the array of prefabricated transistors or gates.

GDSII A design data format used to generate artwork for mask-making. GDSII Stream Format (Gerber Data Structure v. II), or Graphic Design Station II. *(G-D-S-2)*

Germanium Common semiconductor material used to make ICs.

Giga As in gigabyte, one billion of something.

Glitch An input transition or voltage spike that occurs in a time period that is shorter than the delay through the affected logic element. Glitches can propagate to primary outputs, causing functional failure.

Global routing Defining the interconnect paths on a chip between major blocks.

GND Negative supply voltage/ground. *(ground)*

Go, no-go tests Refers to chip tests that decide if a chip works or fails.

GPS Global positioning system—satellite-based worldwide location system.

Ground bounce A switching transient caused by a large number of high-drive IC outputs switching simultaneously. (See also **SSO**.)

GUI Graphical user interface. *(goo-ey)*

Hand-off The passing of information (or responsibility) from one group to another.

Hardware Equipment made of hard components, e.g., chips, boards, etc.

Hardware design The process of developing the hardware part of a system.

Hardware emulation — The process of creating a hardware model of the hardware chip design, usually using FPGAs to emulate the IC design.

Hardware simulation — The process of testing the hardware design.

Hardware/software integration — The process of testing (new) software on (new) hardware.

Hardware/software co-design — The process of developing new hardware and software concurrently (in parallel, in contrast to sequentially).

HDL — Hardware description language; a high-level behavioral abstract of a design, defined in a software algorithm. HDLs usually include behavioral, RTL, and gate-level descriptions.

Henry — Measure of inductance (also milli-henries).

Hertz — Measure of frequency (also kilohertz, megahertz, gigahertz).

Hierarchy — A sequence of something from a higher level to lower levels, e.g., a physical hierarchy for electronic products would include a PC board, ICs, logic blocks, cells, and transistors.

HLL — High-level language—any programming language enabling the programmer to work at a higher level of abstraction.

Hot electron effect — Reliability problem with high electric fields at small dimensions which can damage a transistor.

Hot spot(s) — Localized locations on a chip heated due to heavy current or high switching rates.

I/O — Input/output—the signals and connections into and out of a block or design.

IC — Integrated circuit—tiny semiconductor circuit with both electronic devices (transistors, resistors, capacitors) and interconnect wiring integrated in or on the chip. (See also **Chip**, **Die**.)

IC architecture — The arrangement (or structure) of the transistors (or gates) on the chip.

ICCAD — International Conference on Computer-Aided Design. *(I-cad)*

IEC — International Electronics Consortium.

IEEE — Institute of Electrical and Electronic Engineers. *(I-triple-E)*

IEEE-CAS — IEEE Circuits and Systems Society. *(I-triple-E-cas)*

IDM — Integrated device manufacturer (usually an IC manufacturer).

Inductance — Measure of electrical inertia of a current flow (measured in henries, or millihenries).

Infant mortality	Device failures during the early period of its life.
In-house	Doing work inside the company.
Instance	An occurrence of a cell in a schematic, netlist, or layout.
Insulator	A material such as glass or plastic that does not allow electrons to flow easily.
Interface	General term for the connection between any two things (physical contact, input/output, serial/parallel, RF, etc.).
Interconnect	The wire or wiring used to electrically connect circuit components or gates.
Interoperability	Refers to tools which can exchange data.
IP	Intellectual property—blocks of predesigned blocks (also cores, macros, virtual components). Hard IP—physical design fixed; Soft IP—physical design not fixed: Firm IP—FPGA design block. Also, Internet Protocol—message handshaking over the Internet.
Implant, ion implant	Method of embedding ions (atoms) of one material into another. (See also **Diffusion**, **Doping**.)
ISQED	International Symposium on Quality in Electronic Design.
ISSCC	International Solid-State Circuits Conference.
IT	Information Technology group which administers the company computers and communications facilities.
ITC	International Test Conference.
Iteration	The process of doing something, testing it, and re-doing it. IC design often involves many iterations.
ITRS	International Technology Roadmap for Semiconductors—tri-annual ten-year prediction of semiconductor process parameters.
ITS	Intelligent transport systems.
JEDEC	Joint Electronic Device Engineering Council—many international electronic standards. *(jed-eck)*
JTAG	Joint Test Action Group—IEEE standard for scan testing. *(J-tag)*
Layout	See **Physical design**.
Learning curve	Graph of proficiency versus time in learning a new subject. EDA tools can take several months to learn.
LED	Light-emitting diode. *(L-E-D or led)*
Leverage	Any mechanism that provides a large output from a small effort.

Library Collection of design elements, with logical, physical, etc., views.

Licensing model The particular terms of the business contract used to sell or rent EDA tools, licensing options (e.g., location, use by, duration, charges, portability).

Linear A form of analog circuitry.

Linux Low-cost variation of UNIX operating system. *(lin-ux)*

LINT An early "lexical interpreter" syntax and error checking program developed for C. Now lint-like programs check most design languages. *(lint)*

Load The resistance and/or capacitance that the inputs of one or more devices present to the output of a driving device to which it is connected.

Logic In electronics, refers to electrical circuits that perform symbolic functions (e.g., AND, OR, NOR) where the information is either true or false.

Logic design The activity of selecting and connecting logic circuits to perform a desired function.

LSI Large-scale integration.

LVS EDA tool for layout versus schematic error-checking.

Macro cell/hard macro Also referred to as a core function, a macro is a complex ASIC cell performing some standard function. Hard macros are so-called because their physical layout is fixed in the design rules for which they were originally designed and characterized. (See also **IP**.)

Macro function/cell/ soft macro Unlike hard macros, which are defined at the physical layout level, soft macros are defined at the cell library and netlist level. A soft macro may implement the same electrical function as that of a hard macro, but has no predetermined physical layout.

Manhattan routing Interconnection paths defined at right angles, in contrast to all-angle routing.

Manufacturing test A test used after manufacturing ICs versus tests used during the design phase.

Maze router EDA tool for defining interconnect paths for general wiring.

Mask A glass plate template consisting of clear or opaque areas that respectively allow or block light to shine through. The masks are aligned with existing patterns on the wafer and are used to expose photoresist for the defining of circuit elements and wires.

MCM Multi-chip module. Multiple chips assembled into a single IC package, usually interconnected on miniature PC board.

MEBES Manufacturing electron beam engraving system for lithography photomask production system or its file, or format. *(me-bes)*

Memory IC elements which can store data. RAM is random-access memory. DRAM memory needs continual refresh clocking. SRAM is static RAM. ROM memory has hardwired (permanent) data. PROM memory is (usually one-time) programmable by the user. EPROM is erasable programmable memory (usually by ultraviolet light). EEPROM is electrically erasable EPROM. Volatile memory loses data if the chip loses power. Non-volatile memory saves stored data even without power. ROM, PROM, EPROM, EEPROM, and Flash are non-volatile.

Merging Final IC physical implementation steps of assembling all the cell and device and wiring layers to make the manufacturing files.

Metallization The process of depositing layers of high-conductivity material such as aluminum, used to interconnect circuit elements or cells on a chip.

Metal migration Metal atoms moving (migrating) due to heavy currents in wires. Can cause wire thinning and breaks.

Micron One-millionth of a meter. Submicron—less than a micron.

Microprocessor A CPU or processor on a chip.

Middle-out design A design style which starts with some known block or partition at a mid-level in the hierarchy. Design is done both up and down from there. (See also **Bottom-up** and **Top-down**.)

Mil One-thousandth of an inch.

MIPS Millions of instructions per second. *(mips)*

Mixed mode, mixed signal A design that integrates both digital and analog circuits on the same device, or in the same analysis. Also type of simulator.

Model, modeling A software functional or behavioral representation of a physical device, including its timing characteristics. Modeling may be of different types (e.g., performance, power) and at different levels (e.g., system, logic, or physical levels).

Module generator Software tools which create functional blocks according to a set of input parameters.

Moore's Law	Doubling of IC transistor density every 18-24 mo. Observation made by Gordon Moore of Intel in late 1960s. Measure of increased semiconductor performance and lower cost.
MOS	Metal-oxide semiconductor. A transistor formed with a P or N source and drain, and a metal gate over an insulator forms a N or P channel when a charge is put on the metal gate. PMOS has a P-channel, and NMOS has an N-channel, and CMOS uses both kinds in a circuit, one on when the other is off in a complementary action. *(mos, or M-O-S)*
MOSFET	Metal-oxide semiconductor field-effect transistor. *(mos-fet)*
MPP	Massively parallel processing (an array of 64 or more processors).
NAND	A logic gate that gives a "0" output if all inputs are "1"s. *(nand)*
Nano	One-billionth of something.
Net	Short for network; a circuit path.
Net delay	The (largest) signal delay occurring along a interconnect net.
Netlist	A hardware design description list of a design's transistors, cells or gates, and their connectivity.
NIST	National Institute of Science and Technology. U.S. Dept. of Commerce research facility and funding group. *(nist)*
Node	A terminal or point in a circuit element or any branch of a net.
Non-proprietary	Refers to something not owned by a single company or vendor.
NOR gate	A logic gate that gives a "0" output if one or more inputs are "1". *(noar)*
NPU	Network Processor Unit.
NRE	Non-recurring engineering cost—the IC development cost.
OAC	Open-Access Coalition. User-backed group with data model and application programming interface for any database.
Observe	General requirement of being able to see all the logic outputs during test.
OCP	Open core Protocol—international partnership to create a core connection standard.
OEM	Original equipment manufacturer.
Ohms	Measure of electrical resistance (also milli-ohms, kilo-ohms, mega-ohms).

OMF, OMI	Open Model Forum—a group standardizing interfaces between simulators and reusable models. Open Model Interface (OMI).
On-chip bus	Electrical bus (parallel group of signal wires) connecting blocks on the chip.
OOP	Object-oriented programming—software style that uses well-defined modular chunks of code. *(oop)*
OPC	Optical proximity correction, EDA mask features used to achieve higher resolution when IC manufacturing line widths are smaller than the photolithography light wavelength.
Open standard	A standard which is open to any user, not just customers.
Optimization	A manual or mathematical procedure to achieve the most efficient design.
OR gate	Logic gate that gives a "1" output if one or more inputs is a "1". *(or)*
OS, Operating system	Foundation software which supports application software (e.g., UNIX, Linux, Windows).
OSCI	Open Systems C Initiative—promoting an open source C-language-based system design language.
Outsourcing	Contracting for work or materials outside the company.
OVI	Open Verilog International (merged into Accellera).
Oxide	Refers to silicon dioxide, a layer grown over silicon to protect and insulate it and define areas for devices such as resistors and transistors.
P&R	Place and route (physical layout).
Pad(s)	A metallized area on the chip that is used to connect the I/O circuitry to the package or substrate, with wires or solder balls.
Pad-limited	Pad-limited designs result when the number of input, output, power, and ground pads in the periphery of the chip determine the minimum die size.
PAL	Programmable array of logic. *(pal)*
Parameter	Any variable or aspect of a design that affects the design function or performance (e.g., size, speed, power, etc.).
Parasitic	A resistive, capacitive, inductive, or semiconductor effect due to the close proximity of wires or other elements.

Parasitic extraction	EDA tools which translate IC layout data into circuit models and account for the parasitic parameters.
Partitioning	The process of allocating functions to specific blocks of hardware or software.
Passivation	A layer of material, typically glass, deposited over a completed IC to stabilize its surface and provide protection from contamination.
Patch	A fix or repair to software code.
Patterns	Shapes used in mask-making. Also test vectors, the input/output patterns of 1s and 0s used for testing of designs.
Perl	Programming language used as script for EDA design flows.
PC	Personal computer.
PCB, PC board	Printed circuit board. Plastic/fiberglass board with layers of printed wiring to hold and interconnect electronic devices.
PGA	Pin-grid array.
Pico	One-trillionth of something.
PSM	Phase shifting mask—technique for finer photolithography.
Photolithography	IC fabrication process using stencil-like masks and light to create patterns on a silicon wafer surface.
Photoresist	Light-sensitive material used to transfer mask patterns onto wafer surface. After development and rinsing, a pattern of opaque areas and clear areas remains.
Physical design, physical layout	The physical implementation of an integrated circuit layout in terms of the geometric elements comprising transistors, cells, and blocks, as well as their placement and routing.
Pins	IC circuit package connections to PC board.
PLA	Programmable logic array.
Placement	The physical locating of a transistor, cell, or block on a chip.
Platform	A platform chip provides a base of functions targeted for a specific application target. Portions of the chip are intended to be tailored for or by the customer.
PLD	Programmable logic device—generic term, includes FPGAs, EEPROMs, CPLD (complex PLD), etc.
PLL	Pre-linked library file name extension.
PLL	Phase locked loop—clock and RF component.

Plots, plotter	Large plotter machines that draw detail layout drawings of IC layout.
Point tools	Individual EDA tools—not part of an interoperable group of tools.
Polycrystalline silicon	Silicon composed of many single crystals having a random arrangement. Also known as polysilicon, or simply poly. Polysilicon is often used as an intracell interconnect material and also performs well as a capacitor plate.
Polygon(s)	The graphical patterns that define each transistor and interconnect area for the masks used to make an IC.
Porting	Moving a design from one process to another. (See also **Retargeting**.)
Ports	Design description objects that model input, output, and bidirectional interfaces.
Power analysis	EDA tools which calculate the power and current all over the chip.
Power buses	Heavy interconnect wires to distribute power on chip.
Power routing	Interconnect paths specifically drawn for power distribution.
Primitive	A low-level instance of a cell function such as a gate.
Process	Semiconductor manufacturing technology—all the activity and equipment required to implement ICs. Measured by smallest feature size attainable by given process.
Processor	The part of a computer that does the calculation or computing. (See also **CPU**.)
Production test	Final tests done to ensure that a product was made correctly before shipping.
Production tool	Tool that works, is well-documented, exhaustively tested, user-friendly, and stable (in contrast to prototype or university tools).
Productivity Gap	Apparent gap on a graph between the increasing manufacturing capability in number of transistors per IC, and the more gradual increase in EDA-supported design capability in number of transistors per IC.
Programmers	People who write computer programs. Also, hardware boxes that load programs into programmable chips such as FPGAs.
PROM	Programmable read-only memory. (See also **Memory**.) *(prom)*
Properties	In formal verification, the description of the logical intent. Property-checking tools verify conformance to properties.

Proprietary Refers to ownership by a single company or vendor.

Prototype An original design or first operating model intended for evaluation of the form, fit, and function for a particular application. Prototype ASIC devices are representative of the final production units.

PVT Power, voltage, and temperature.

Qualified Parts List List of devices, IP, or suppliers which have been qualified for the designer to use on a given project (QPL).

RC R and C, resistance and capacitance, related because their product R*C affects the time delay of the device or interconnect.

Race condition A situation which occurs if a data signal arrives too early or late with respect to the clock. The signal may not get stored in a flip-flop or register.

R & D Research and development.

Race condition The condition that results when a signal is propagated through two or more logic or memory elements in the same clock period, thus violating the timing requirements for proper circuit operation. Also called a timing hazard.

RAM Random access memory. (See also **SRAM**, **DRAM**.) *(ram)*

Random logic Logic outside the major functional blocks of a design. Also called "glue" logic. Often the control logic in a design.

Rapid prototyping A way of building a quick hardware model of a system using pre-existing computers and FPGAs.

Real-time Referring to hardware and software that must perform their critical function and respond within a time window constrained by real events.

Reconfiguration The changing of all or part of an FPGA.

Register A row of flip-flops which can store a byte or word of data in a computer, used for fast access/storage of temporary data.

Release Event of releasing hardware or software (new or major revision / update) to users.

Resistance Measure of the difficulty which an electrical current has flowing through a material (measured in ohms).

Re-size The process of adjusting the transistor feature sizes to a new process. (See also **Porting**.)

Re-spin A redo of a new silicon chip due to a design error.

Re-targeting Porting an existing design from one process technology to another.

Re-use The use of previous designs (re-usable logic blocks or any kind of intellectual property). "Re-use without rework" uses the prior design as is. "Re-use with rework" modifies the prior design and requires re-verification.

RF Radio frequency, high-frequency radio waves or the circuitry used to work with high-frequency signals.

RISC Reduced instruction set computer. *(risk)*

ROI Return on investment ($ spent versus $ made).

ROM Read-only memory. (See also **Memory**.) *(rom)*

Routing The interconnection paths on a chip. An EDA router tool defines the paths. (See also **Channel**, **Detail**, **Global**, **Maze Routers**.)

RTL Register transfer-level design (moving data between registers).

RTOS Real-time operating system. *(R-tos)*

RTP Research Triangle Park, Real-Time Programming.

Run-time control Refers to the process or tools to guide and control the design flow including change control, tracking related files, etc.

SAR Segmentation and re-assembly communications circuit. *(sar)*

SC Standard cell.

Scan See **Boundary scan**.

Scheme Programming language used as script for EDA design flow.

Schematic, diagram A picture or diagram of a logic or electrical circuit using symbols and wires. Also schematic entry or capture.

SDF Standard delay format, a uniform way of representing delay data.

Sea of gates A gate array architecture that features a continuous array of transistors. Sea-of-gates arrays utilize unused transistor sites for routing, as they have no dedicated routing channels.

Semantics The meaning of a word or sentence.

Semiconductor A material like silicon that conducts electricity poorly, but can be enhanced to conduct more readily.

Semicustom Refers to any ASIC methodology that utilizes prefabricated and characterized circuit elements, thus requiring only element interconnect for customization (gate arrays, standard cells).

Sematech International industry consortium developing near-term advances in semiconductor manufacturing.

SEMI Semiconductor Equipment Manufacturers International. *(semi)*

Server (s) High-performance, high-storage computer(s) shared by users.

Shrink Refers to process shrinks, where only a small change in feature size or wire size occurs.

SI Signal integrity; how well a signal is defined as a "1" or "0".

SIA Semiconductor Industry Association.

SIGDA Special Interest Group on Design Automation (ACM). *(sig-D-A)* Software group for EDA professionals.

Signal Electrical transfer of information between electrical components. Signals going into a component are inputs, and those coming out are outputs.

Signal delay The time required for a change in a signal (0-1 or 1-0) to pass through a gate or interconnect path.

Signal integrity (SI) Signal integrity analysis measures how well a signal travels from one logic gate to another. Many sources of noise affect the signal integrity.

Silicon Common semiconductor material used to make ICs. P-type silicon is doped to have more holes, N-type silicon is doped to have more electrons.

Silicon dioxide, SiO_2 A insulating compound of silicon and oxygen, used to isolate active transistors and interconnect. (Quartz and sand are SiO_2.) *(s-i-oh-2)*

Simulation, simulator A program that models the operation of individual components or gates and their interconnect and timing. Used to test a design written in a hardware description language (e.g., Verilog).

SIP Silicon intellectual property (see also **IP**, **Core**, **Macro**, **SLB**, etc.) also single inline package, system-in-package. *(sip)*

SI2 Silicon Initiative, Inc., a group related to the ASIC Council, leading OAC effort.

SLDL System-level description language.

SNUG Synopsys User Group, independent EDA user community. *(snug)*

SRAM Static RAM. *(S-ram)*

Solder Soft metal mix of tin and lead used to attach wires, ICs. *(soder)*

SoC System-on-chip—all or most of a system's electronics on a chip. It usually includes one or more processors, memories, logic, and I/O interfaces.

Software Programs of instructions that run on computers.

Software design The process of developing the software part of a system.

Software languages Shorthand languages that enable programmers to write programs more efficiently.

Software verification The process of testing the software design.

SOG Sea-of-gates.

SOI Silicon-on-insulator—a manufacturing approach to improve transistor performance and isolation.

Source Current source in a MOS transistor.

Source code Software program code written in original "high-level" user language (versus object code—software translated into computer-level code).

SPARC Scalable processor architecture. *(spark)*

Specification A document describing the design requirements (performance, size, functions, power, etc.).

SPICE Simulation program with integrated circuit emphasis. *(spice)*

SRAM Static random access memory. *(S-ram)*

SSO Simultaneously switching output. (See also **Ground bounce**.)

STA Static timing analyzer. A program that calculates logic timing without the need for test patterns.

Standard cell, SC A primitive functional element such as a gate or latch that is characterized by fixed physical and electrical characteristics.

Also, the IC architecture that utilizes standard cells.

STAR IP IP blocks which are unique, leading-edge designs. (*star*)

State, state diagram The collective information stored in each flip-flop or memory element in the machine at a given time—it is the machine state at that time. Also, state transition diagram.

Static An unclocked mode or condition.

Static electricity Accumulation of electrons on insulators. Can reach high voltage.

Switching noise Noise generated by rapidly switching transistors.

Symbol	A graphical representation of a cell featuring its bounding box and I/O ports. Symbols are used primarily in schematic editing.
Synchronous	A design technique whereby the performance of operations is controlled by regular clock intervals.
Syntax	Sentence structure—arrangement of words.
Synthesis	The translation of a high-level design (RTL) description consisting of state transition machines, truth tables, and/or Boolean equations into a process-specific gate-level logic implementation.
SYSCLK	System clock—primary clock of the design. *(sys-clock)*
Tapeout	Final physical design file to send to mask shop for IC fab.
TAP	Test access port (IEEE standard). *(tap)*
TAT	Turn-around time. *(tat)*
TCAD	Technology computer-aided design. *(T-cad)*
Tcl	Test Command Language. *(tickle)*
TCP/IP	Transmission control protocol/Internet protocol.
TDL	Timing-driven layout.
Test	A procedure used to find out if something works or not. Usually a stimulus is input to the device under test, and the output is observed and compared to the expected output.
Testability	A measure of how testable a design is.
Test bench	A platform or model of the external world in which the device under test is used. The test bench facilitates testing.
Test insertion	The manual or automatic insertion of test logic such as test points and boundary-scan registers into a design.
Test patterns, vectors	Set of input data used to stimulate a device or model for testing.
Third-party company	A company supplying services or materials that is not the primary product manufacturer or user.
Throughput analysis	The study of how fast data will travel through a system under varying conditions.
Timing-driven design	Designing with focus on optimizing the chip timing (versus the chip size or cost or power).
Timing closure	The procedure of finding timing (delay) problems, fixing them, and rechecking the design, until all problems are resolved.

Timing diagram	Graphical diagram of time relationship of signals in an electronic design.
TLA	Acronym for "Three-Letter Acronym."
Top-down design	Design style starting at the highest level of abstraction and sequentially designing the lower levels. (See also **Bottom-up** and **Middle-out**.)
Tool suite	A group of interoperable tools that interface or share data with each other. (See also **Point tools**.)
Transformer	Device which converts AC voltage higher or lower.
Transmission line	Conductor or wire for carrying high-frequency signals.
Transistor	Semiconductor amplifier or switching device. Basic element of integrated circuits.
Tree router	EDA tool for defining tree-shaped paths for clocks and power.
Troubleshooting	A procedure for locating the cause(s) of a problem (bug).
TSMC	Taiwan Semiconductor Manufacturing Corporation—foundry.
TTL	Transistor-transistor logic.
TTM	Time to market.
Unit delay simulation	Simplified form of timing simulation where every gate delay is one unit, by default.
UNIX	UNiversal Interactive eXecutive, popular operating system for EDA tools. *(U-nix)*
Validation	The process of ensuring that the design behaves as the customer expected. (See also **Verification**.)
VC	Virtual component (*aka* core, macro, IP block), venture capitalist.
VCD	Value change dump. (Simulator output.)
VCO	Voltage-controlled oscillator.
VCC, VDD	Positive supply voltage, hot lead, power.
VDSM	Very deep submicron (~ below 90 nanometers).
Verification	The process of ensuring that the design behaves as the designer intended. (See also **Validation**.)
Verilog	A text-based description of digital logic, standardized under IEEE and Accellera. It describes structure and behavior.

VHDL	An acronym for VHSIC Hardware Description Language; a standard, technology-independent design description language, standardized under IEEE and Accellera. It describes behavior, RTL and gate-level logic descriptions.
VHSIC	Very high-speed integrated circuit—old mil program. *(vis-ick)*
Via	Layer-to-layer wiring connection on PC boards and ICs.
View	One of several possible representations of a design.
Virtual prototype	A software model of hardware in development used to support early testing of new application programs intended to run on that hardware.
VLSI	Very large-scale integration.
VLSIC	Very large-scale integrated circuit.
Voltage	Measure of electrical pressure that drives electrons through conductors (also millivolts, microvolts).
Voltage sensitivity	A measure of the effect which variations in voltage across a chip have on the chip circuits.
VSI	Virtual socket interface.
VSIA	Virtual Socket Interface Alliance, system-on-chip standards.
VSS	Negative supply voltage, also ground, grd, gnd, return.
Wafer	The silicon disc on which multiple IC chips are made.
Wafer fab	An IC fabrication factory.
Wafer test	Tests which are run on individual IC chips while still part of the wafer.
Waves	Electromagnetic waves, radiated by rapidly changing electrical currents. (Radio, X-ray, and infrared waves are examples.)
White box	Implementing or testing an IP block whose internal details are accessible to the designer.
Word	4-, 8-, 16-, 32-, 64- (etc.) bit chunk of data, usually matching the width of the data a computer is designed for. (See also **Byte**.)
Workstation	Individual computer equipment for IC designer (usually high-performance, high-storage, fast LAN capability).
WSTS	World Trade Semiconductor Trade Statistics—non-profit semiconductor sales statistics organization.
X-Initiative	Industry partnership to promote 45-degree routing.

Yield The number of units surviving screening operations.

Yield improvement Techniques for getting more working chips from a wafer.

Zap Common term for burning out an electrical component. (See also **ESD**.)

Index

A

B

C

D

E

F

G

H

I

R

S

T

U

V

Z

About the Author

Mark D. Birnbaum holds a BSEE from Princeton University and an MSEE from Northeastern University. He held senior positions in engineering design, consulting, technology partnerships, and technical marketing with both commercial and aerospace firms.

In addition, he managed engineering, research and software groups at many large and small companies, including National Semiconductor Corp., Fujitsu Microelectronics, Inc., and Cadence Design Systems. Along the way, he led two standards groups, wrote two books, and developed and taught several microelectronics and EDA classes.

Mr. Birnbaum has been an EDA user, manager, developer, and vendor. His diverse technology experience includes EDA, PCB, MCM, IC, IC packaging, ASIC, SoC, FPGA, A/D, D/A, low power, IP evaluation and electronics quality.

Currently, Mr. Birnbaum does technical writing and consulting, and is known as "An Engineer Who Can Write!" His writing experience includes white papers, conference papers, business plans, magazine articles, customer success stories, executive presentations, and DARPA research proposals. He has had articles and papers published in many magazines and conferences, including EETimes, Electronic Design, ISSCC, ISQED, and DATE.

When not traveling or walking on the beach in Carlsbad, CA, Mr. Birnbaum enjoys reading, teaching, and writing. He can be reached at mdbirnbaum@ieee.org